# McGraw-Hill My Math

## Interactive Guide

Grade 1

ConnectED.mcgraw-hill.com

Copyright © 2014 McGraw-Hill Education

All rights reserved. No part of this publication may be reproduced or distributed in any form or by any means, or stored in a database or retrieval system, without the prior written consent of McGraw-Hill Education, including, but not limited to, network storage or transmission, or broadcast for distance learning.

**STEM** McGraw-Hill is committed to providing instructional materials in Science, Technology, Engineering, and Mathematics (STEM) that give all students a solid foundation, one that prepares them for college and careers in the 21st century.

Send all inquiries to:
McGraw-Hill Education
8787 Orion Place
Columbus, OH 43240

Selections from:
ISBN: 978-0-02-132705-8 *(Grade 1 Student Edition)*
MHID: 0-02-132705-X *(Grade 1 Student Edition)*
ISBN: 978-0-02-130895-8 *(Grade 1 Teacher Edition)*
MHID: 0-02-130895-0 *(Grade 1 Teacher Edition)*

Printed in the United States of America.

Visual Kinesthetic Vocabulary® is a registered trademark of Dinah-Might Adventures, LP.

7 8 9 LOV 20 19 18 17

# Contents

## Chapter 1 Addition Concepts

Mathematical Practice 6/Inquiry . . . . 1
Lesson 1 Addition Stories . . . . . . . 2
Lesson 2 Model Addition . . . . . . . 3
Lesson 3 Addition Number
    Sentences . . . . . . . . . . . . . . . 4
Lesson 4 Add 0 . . . . . . . . . . . . . 5
Lesson 5 Vertical Addition . . . . . . 6
Lesson 6 Problem Solving Strategy:
    Write a Number Sentence . . . . . 7
Lesson 7 Ways to Make 4 and 5 . . . 8
Lesson 8 Ways to Make 6 and 7 . . . 9
Lesson 9 Ways to Make 8 . . . . . . 10
Lesson 10 Ways to Make 9 . . . . . 11
Lesson 11 Ways to Make 10 . . . . 12
Lesson 12 Find Missing Parts of 10 . . . 13
Lesson 13 True and False
    Statements . . . . . . . . . . . . . 14

## Chapter 2 Subtraction Concepts

Mathematical Practice 6/Inquiry . . . 15
Lesson 1 Subtraction Stories . . . . . 16
Lesson 2 Model Subtraction . . . . . 17
Lesson 3 Subtraction Number
    Sentences . . . . . . . . . . . . . 18
Lesson 4 Subtract 0 and All . . . . . 19
Lesson 5 Vertical Subtraction . . . . 20
Lesson 6 Problem Solving Strategy:
    Draw a Diagram . . . . . . . . . 21
Lesson 7 Compare Groups . . . . . 22
Lesson 8 Subtract from 4 and 5 . . . 23
Lesson 9 Subtract from 6 and 7 . . . 24
Lesson 10 Subtract from 8 . . . . . 25
Lesson 11 Subtract from 9 . . . . . 26
Lesson 12 Subtract from 10 . . . . 27
Lesson 13 Relate Addition and
    Subtraction . . . . . . . . . . . . 28
Lesson 14 True and False
    Statements . . . . . . . . . . . . 29

## Chapter 3 Addition Strategies to 20

Mathematical Practice 1/Inquiry . . . 30
Lesson 1 Count on 1, 2, or 3 . . . . . . 31
Lesson 2 Count On Using Pennies . . . 32
Lesson 3 Use a Number Line to Add . . . . . . 33
Lesson 4 Use Doubles to Add . . . . . 34
Lesson 5 Use Near Doubles to Add . . 35
Lesson 6 Problem Solving Strategy: Act It Out . . . . . . 36
Lesson 7 Make 10 to Add . . . . . . . 37
Lesson 8 Add in Any Order . . . . . . 38
Lesson 9 Add Three Numbers . . . . . 39

## Chapter 4 Subtraction Strategies to 20

Mathematical Practice 1/Inquiry . . . 40
Lesson 1 Count Back 1, 2, or 3 . . . . 41
Lesson 2 Use a Number Line to Subtract . . . . . . 42
Lesson 3 Use Doubles to Subtract . . . 43
Lesson 4 Problem Solving Strategy: Write a Number Sentence . . . . . 44
Lesson 5 Make 10 to Subtract . . . . . 45
Lesson 6 Use Related Facts to Add and Subtract . . . . . . 46
Lesson 7 Fact Families . . . . . . . 47
Lesson 8 Missing Addends . . . . . . 48

## Chapter 5 Place Value

Mathematical Practice 2/Inquiry . . . 49
Lesson 1 Numbers 11 to 19 . . . . . . 50
Lesson 2 Tens . . . . . . 51
Lesson 3 Count by Tens Using Dimes . . . . . . 52
Lesson 4 Ten and Some More . . . . . 53
Lesson 5 Tens and Ones . . . . . . 54
Lesson 6 Problem Solving Strategy: Make a Table . . . . . . 55
Lesson 7 Numbers to 100 . . . . . . 56
Lesson 8 Ten More, Ten Less . . . . . 57
Lesson 9 Count by Fives Using Nickels . . . . . . 58
Lesson 10 Use Models to Compare Numbers . . . . . . 59
Lesson 11 Use Symbols to Compare Numbers . . . . . . 60
Lesson 12 Numbers to 120 . . . . . . 61
Lesson 13 Count to 120 . . . . . . 62
Lesson 14 Read and Write Numbers to 120 . . . . . . 63

## Chapter 6 Two Digit Addition and Subtraction

| | |
|---|---|
| Mathematical Practice 4/Inquiry | 64 |
| Lesson 1 Add Tens | 65 |
| Lesson 2 Count on Tens and Ones | 66 |
| Lesson 3 Add Tens and Ones | 67 |
| Lesson 4 Problem Solving Strategy Guess, Check, and Revise | 68 |
| Lesson 5 Add Tens and Ones with Regrouping | 69 |
| Lesson 6 Subtract Tens | 70 |
| Lesson 7 Count Back by Tens | 71 |
| Lesson 8 Relate Addition and Subtraction of Tens | 72 |

## Chapter 7 Organize and Use Graphs

| | |
|---|---|
| Mathematical Practice 4/Inquiry | 73 |
| Lesson 1 Tally Charts | 74 |
| Lesson 2 Problem Solving Strategy: Make a Table | 75 |
| Lesson 3 Make Picture Graphs | 76 |
| Lesson 4 Read Picture Graphs | 77 |
| Lesson 5 Make Bar Graphs | 78 |
| Lesson 6 Read Bar Graphs | 79 |

## Chapter 8 Measurement and Time

| | |
|---|---|
| Mathematical Practice 5/Inquiry | 80 |
| Lesson 1 Compare Lengths | 81 |
| Lesson 2 Compare and Order Lengths | 82 |
| Lesson 3 Nonstandard Units of Length | 83 |
| Lesson 4 Problem Solving Strategy: Guess, Check, and Revise | 84 |
| Lesson 5 Time to the Hour: Analog | 85 |
| Lesson 6 Time to the Hour: Digital | 86 |
| Lesson 7 Time to the Half Hour: Analog | 87 |
| Lesson 8 Time to the Half Hour: Digital | 88 |
| Lesson 9 Time to the Hour and Half Hour | 89 |

## Chapter 9 Two-Dimensional Shapes and Equal Shares

| | |
|---|---|
| Mathematical Practice 7/Inquiry | 90 |
| Lesson 1 Squares and Rectangles | 91 |
| Lesson 2 Triangles and Trapezoids | 92 |
| Lesson 3 Circles | 93 |
| Lesson 4 Compare Shapes | 94 |
| Lesson 5 Composite Shapes | 95 |
| Lesson 6 More Composite Shapes | 96 |
| Lesson 7 Problem Solving Strategy: Use Logical Reasoning | 97 |
| Lesson 8 Equal Parts | 98 |
| Lesson 9 Halves | 99 |
| Lesson 10 Quarters and Fourths | 100 |

## Chapter 10 Three-Dimensional Shapes

Mathematical Practice 7/Inquiry . . . 101

Lesson 1 Cubes and Prisms . . . . . 102

Lesson 2 Cones and Cylinders . . . . 103

Lesson 3 Problem Solving Strategy: Look for a Pattern . . . . . . . . 104

Lesson 4 Combine Three-Dimensional Shapes . . . . . . . . . . . . . 105

Visual Kinesthetic Vocabulary© . . . VKV1

NAME _____  DATE _____

# Chapter 1 Addition Concepts
*Inquiry of the Essential Question:*

## How do I add numbers?

| ● Part | ○ Part |
|---|---|
| ● ● ● ● | ○ |
| Whole ||
| ● ● ● ● ○ ||

I see ...

I think ...

I know ...

| Part | Part |
|---|---|
| 3 | 4 |
| Whole ||
| 7 ||

I see ...

I think ...

I know ...

● ● ● + ● ● = ● ● ● ● ●

I see ...

I think ...

I know ...

Questions I have...

_____

_ _ _ _ _ _ _ _ _ _ _ _ _ _ _ _ _ _ _ _ _ _ _ _ _

_____

_ _ _ _ _ _ _ _ _ _ _ _ _ _ _ _ _ _ _ _ _ _ _ _ _

**Teacher Directions:** Read the Essential Question for students. Have students echo read. Direct students to describe their observations, inferences, and prior knowledge of each math example. Encourage students to write or draw additional questions they may have. Then have students share their thinking/questions with a peer.

Grade 1 • Chapter 1 *Addition Concepts*  1

NAME _____ DATE _____

# Lesson 1 Four-Square Vocabulary
## Addition Stories

Trace the word. Write the definition for *number*. Write what the word means, draw a picture, and write your own sentence using the word.

| **Definition** | **My Own Words** |
|---|---|
| | |
| **My Picture** | **My Sentence** |

(center: number)

**Teacher Directions:** Provide a description, explanation, or example of the new term using images or real objects. Have students trace the word and then use the Glossary to write the definition. Direct students to write a definition in their own words and draw a picture representing their math term. Have students write a sentence using the term. Then encourage students to read their sentence to a peer.

NAME _____    DATE _____

# Lesson 2 Note Taking
## *Model Addition*

Read the question. Write words you need help with. Use your lesson and the word bank to write your Cornell notes. Write or draw math examples to explain your thinking.

| **Building on the Essential Question** How can I model addition? | **Notes:** |
| --- | --- |
| | **Word Bank**: objects, part, add, whole |
| | A _____ is a set of objects I want to join. |
| | When I add, I want to find the _____. |
| **Words I need help with:** | The whole is all of the _____ in a group. |
| | I must _____ the parts to find the whole. |
| **My Math Examples:** | 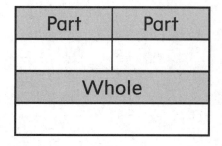 |

**Teacher Directions:** Read the Building on the Essential Question and have students list words/phrases they need assistance with. Provide descriptions, explanations, or examples of the terms using images or real objects. Read each sentence frame and have students fill in the appropriate terms from the word bank. Have students read their notes aloud. Direct students to draw a picture representing the question. Then encourage students to describe their picture to a peer.

Grade 1 • Chapter 1 Addition Concepts   3

# Lesson 3 Word Identification
## Addition Number Sentences

Match each term to a symbol or number.

sum                    $+$

plus                   $=$

equals     $2 + 1 = ③$ ←

---

Write the correct term from above for each sentence on the blank lines.

**addition number sentence**

$2 + 1 = 3$

Two _____ one _____ three.

The _____ is three.

**Teacher Directions:** Review the terms using manipulatives, such as counters or pennies. Have students say each word and then draw a line to match the word to its symbol or number. Direct students say the addition number sentence and then write the corresponding terms in the sentences. Encourage students to read the sentences to a peer.

NAME _____ DATE _____

# Lesson 4 Word Web
## Add 0

Use the word web to show examples of zero.

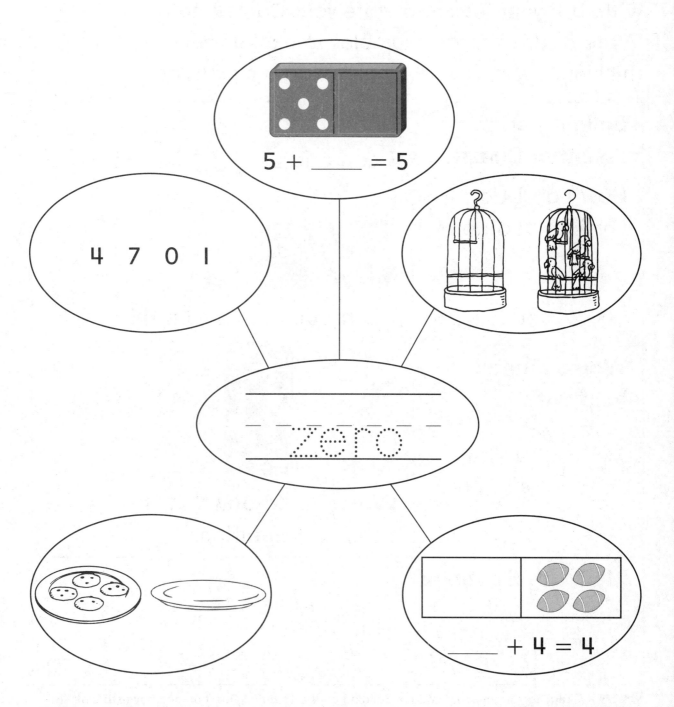

 **Teacher Directions:** Provide a description, explanation, or example of the new term using images or real objects. Have students say the letters aloud as they trace the math term. Direct students to circle the example of zero in each picture, and have them complete the number sentences. Then encourage students to describe their work to a peer.

Grade 1 • Chapter 1 Addition Concepts

# Lesson 5 Note Taking
## Vertical Addition

Read the question. Write words you need help with. Use your lesson to write your Cornell notes. Write or draw math examples to explain your thinking. Share your examples with a classmate.

| **Building on the Essential Question** How can I do vertical addition? | **Notes:** I can add _____ like this: 3 + 4 = 7  |
|---|---|
| **Words I need help with:** | I can add _____ like this: $$\begin{array}{r} 3 \\ +4 \\ \hline 7 \end{array}$$  When I add down, it is _____ addition. |

**My Math Examples:**

**Teacher Directions:** Read the Building on the Essential Question and have students list words/phrases they need assistance with. Provide descriptions, explanations, or examples of the terms using drawings or real objects. Read each sentence frame and have students write the appropriate terms. Have students read their notes aloud. Direct students to draw a picture representing the question. Then encourage students to describe their picture to a peer.

# Lesson 6 Problem Solving

*STRATEGY: Write a Number Sentence*

<u>Underline</u> what you know. (Circle) what you need to find. Write an addition number sentence to solve.

1. **Andrew** saw **3** rabbits.

   **Tia** saw **6** other rabbits.

   How many rabbits did **they** (Andrew and Tia) see **in all**?

rabbit

| Part | Part |
|------|------|
|      |      |
| Whole ||
|      |      |

\_\_\_ + \_\_\_ = \_\_\_

**They** saw \_\_\_ rabbits **in all**.

**Teacher Directions:** Provide a description, explanation, or example of the bold face terms and nouns using images or real objects. Read each sentence and have students echo read. Encourage students to use the part-part-whole mat, complete the addition sentence, and then write their answer in the restated question. Have students read the answer sentence aloud.

NAME _____   DATE _____

# Lesson 7 Sum Identification
## Ways to Make 4 and 5

Circle all the ways to make 5 in the boxes below.

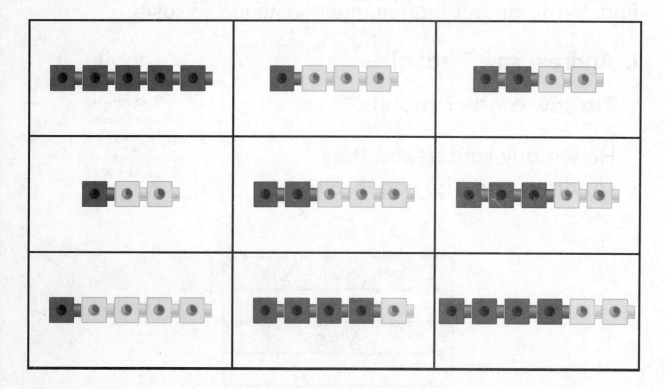

Draw a picture to show one way to make 4.

**Teacher Directions:** Use manipulatives to show various ways to make 4 and 5. Model an addition sentence for each and have students repeat it. Have students circle each example that shows a sum of 5. Then direct students to draw a picture representing a way to make 4. Finally encourage students to describe their picture to a peer.

# Lesson 8 Sum Identification
## *Ways to Make 6 and 7*

Circle all the ways to make 6 in the boxes below.

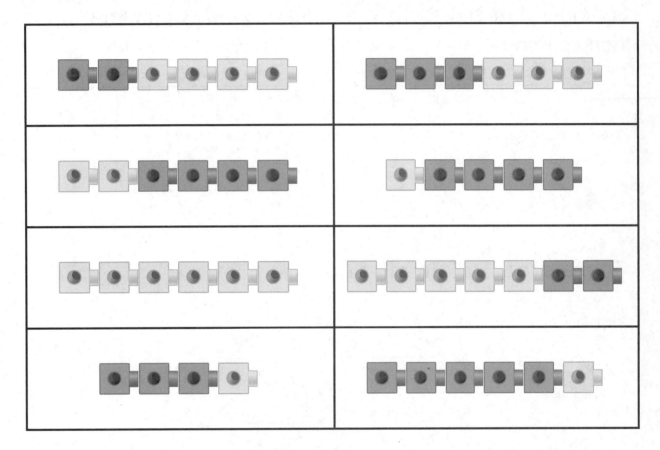

Draw a picture to show one way to make 7.

**Teacher Directions:** Use manipulatives to show various ways to make 6 and 7. Model an addition sentence for each and have students repeat it. Have students circle each example that shows a sum of 6. Then direct students to draw a picture representing a way to make 7. Finally encourage students to describe their picture to a peer.

Grade 1 • Chapter 1 *Addition Concepts*

NAME _____  DATE _____

# Lesson 9 Word Web
## *Ways to Make 8*

Trace the math words. Draw a number story in each rectangle that shows the meaning of *in all*. Complete each sentence.

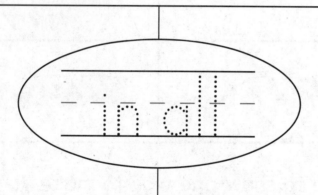

There are _____ in all.

There are _____ in all.

**Teacher Directions:** Provide a description, explanation, or example of the new term using images or real objects. Have students say the letters aloud as they trace the math term. Direct students to draw two picture stories that represent the math term. Have them complete each sentence. Then encourage students to tell their number stories to a peer. Ensure they read their sentences to their peer as well.

NAME _____ DATE _____

# Lesson 10 Vocabulary Definition Map
## *Ways to Make 9*

Use the definition map to write what the math word means and tell what the word is like. Write or draw a math example. Share your examples with a classmate.

**My Math Word:**

add

**What It Means:**

**What It Is Like:**

I can add _____ like this: 3 + 6 = 9.

When I add, I find the ____.

There are many _____ to make a sum of 9.

**My Math Example:**

**Teacher Directions:** Provide a description, explanation, or example of the new term using images or real objects. Have students trace the word then use the lesson or Glossary to define the math term. Direct students to list characteristics and draw a picture representing their math term. Then encourage students to describe their picture to a peer.

Grade 1 • Chapter 1 *Addition Concepts* 11

# Lesson 11 Vocabulary Sentence Frames
## *Ways to Make 10*

The math words in the word bank are for the sentences below. Write the words that fit in each sentence on the blank lines.

| **Word Bank** |
| :---: |
| equals     plus     sum |

1. The answer to an addition problem is the ____.
   
   2 + 4 = 6
   ↑

2. Eight ____ two equals ten.

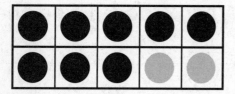

3. One plus nine ____ ten.

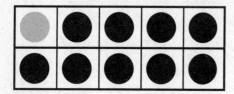

**Teacher Directions:** Provide a description, explanation, or example of the each term using images or real objects. Read each sentence frame and have students echo read. Direct students to write the correct term in each blank. Then encourage students to read each sentence to a peer.

NAME _____   DATE _____

# Lesson 12 Word Identification
## *Find Missing Parts of 10*

Match each term to a picture.

whole

part

missing

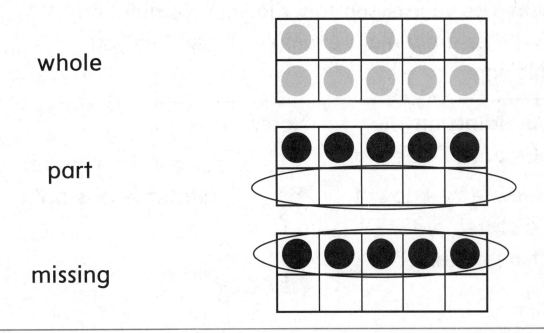

Write the correct term from above for each sentence on the blank lines.

| ● Part | ○ Part |
|--------|--------|
| 7      | _____  |
| **Whole** ||
| 10 ||

The _____ is 10.

7 is a _____.

The _____ part is 3.

**Teacher Directions:** Review the terms using manipulatives, such as counters or pennies. Have students say each word and then draw a line to match the word to its matching picture. Prompt students to describe the part-part-whole mat. Then have them write the corresponding terms in the sentences. Encourage students to read the sentences to a peer.

Grade 1 • Chapter 1 Addition Concepts    13

NAME _____ DATE _____

# Lesson 13 Note Taking
*True and False Statements*

Read the question. Write words you need help with. Use your lesson to write your Cornell notes. Write or draw math examples to explain your thinking.

| **Building on the Essential Question** How do I know if a statement is true or false? | **Notes:** _____ can be true or false. <br> A _____ statement is not a fact. <br><br>      $3 + 1 = 3$   No <br><br> A _____ statement is a fact. <br><br>      $3 + 1 = 4$   Yes <br><br> I should add the numbers to see if the ____ is correct. |
|---|---|
| **Words I need help with:** | |
| **My Math Examples:** | |

 **Teacher Directions:** Read the Building on the Essential Question and have students list words/phrases they need assistance with. Provide descriptions, explanations, or examples of the terms using images or real objects. Read each sentence frame and have students write the appropriate terms. Have students read their notes aloud. Direct students to draw a picture representing the question. Then encourage students to describe their picture to a peer.

14 Grade 1 • Chapter 1 *Addition Concepts*

# Chapter 2 Subtraction Concepts

*Inquiry of the Essential Question:*

## How do I subtract numbers?

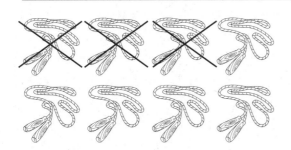

I see ...

I think ...

I know ...

I see ...

I think ...

I know ...

Eight minus three equals five.

8 minus 3 equals 5

$8 - 3 = 5$

Questions I have...

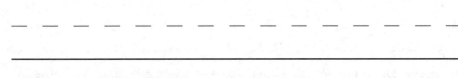

**Teacher Directions:** Read the Essential Question for students. Have students echo read. Direct students to describe their observations, inferences, and prior knowledge of each math example. Encourage students to write or draw additional questions they may have. Then have students share their ideas/questions with a peer.

Grade 1 • Chapter 2 *Subtraction Concepts* **15**

NAME _____ DATE _____

# Lesson 1 Word Web
*Subtraction Stories*

Trace the math words. Draw a number story in each rectangle that shows the meaning of *are left*. Complete the sentences about each story.

_____ **are left.**

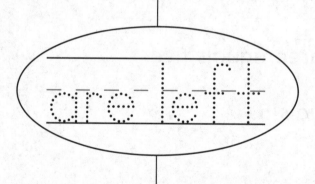

_____ **are left.**

**Teacher Directions:** Provide a description, explanation, or example of the new term using images or real objects. Have students say the letters aloud as they trace the math term. Direct students to draw two picture stories that represent the math term. Have them complete the sentences. Then encourage students to tell their number stories and read their sentences to a peer.

16 Grade 1 • Chapter 2 *Subtraction Concepts*

NAME _____ DATE _____

# Lesson 2 Four-Square Vocabulary
## *Model Subtraction*

Trace the word. Write the definition for *subtract*. Write what the word means, draw a picture, and write your own sentence using the word.

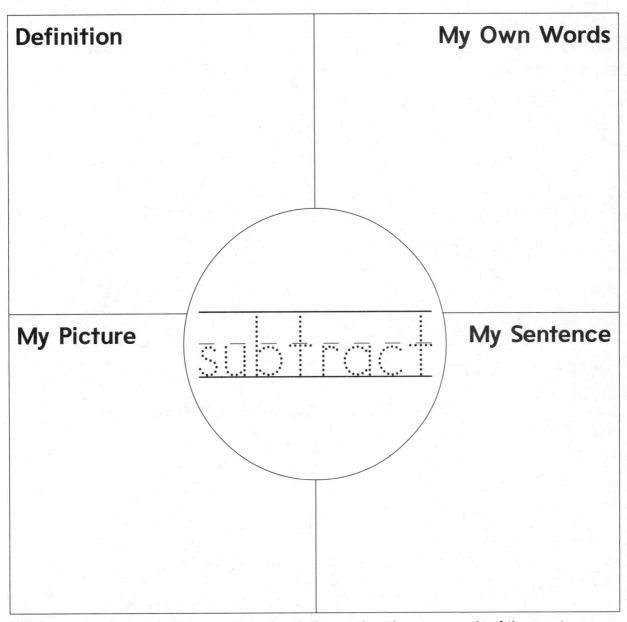

 **Teacher Directions:** Provide a description, explanation, or example of the new term using images or real objects. Have students use the Glossary to write the definition. Direct students to write a definition in their own words and draw a picture representing their math term. Have students write a sentence using the term. Then encourage students to read their sentence to a peer.

Grade 1 • Chapter 2 *Subtraction Concepts*  17

NAME _____ DATE _____

# Lesson 3 Word Identification
## *Subtraction Number Sentences*

Match each term to a symbol or example.

minus                           =


difference                      —


equals              3 − 1 = ②

Write the correct term from above for each sentence on the blank lines.

### subtraction number sentence
3 − 1 = 2

Three _____ one _____ two.

The _____ is two.

**Teacher Directions:** Review the terms using manipulatives, such as counters or pennies. Have students say each word and then draw a line to match the word to its symbol or example. Direct students say the subtraction number sentence and then write the corresponding terms in the sentences. Encourage students to read the sentences to a peer.

18 Grade 1 • Chapter 2 *Subtraction Concepts*

NAME _____ DATE _____

# Lesson 4 Note Taking

*Subtract 0 and All*

Read the question. Write words you need help with. Use your lesson to write your Cornell notes. Write or draw math examples to explain your thinking.

| **Building on the Essential Question** How can I subtract 0? How can I subtract all? | **Notes:** I know that *subtract* means "_____ _____." If I subtract all, I will have ____ left.  ⚾⚾⚾ − ⚾⚾⚾ = 0 |
|---|---|
| **Words I need help with:** | If I subtract 0 from a number, I will have the _____ left. ⚾⚾⚾⚾ − 0 = ⚾⚾⚾⚾ |
| **My Math Examples:** | |

**Teacher Directions:** Read the Building on the Essential Question and have students list words/phrases they need assistance with. Provide descriptions, explanations, or examples of the terms using images or real objects. Read each sentence frame and have students write the appropriate terms. Have students read their notes aloud. Direct students to draw a picture representing the question. Then encourage students to describe their picture to a peer.

NAME _____ DATE _____

# Lesson 5 Concept Web

*Vertical Subtraction*

Write *vertical* in the center oval. Draw lines to match the vertical items to the word *vertical*.

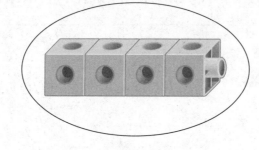

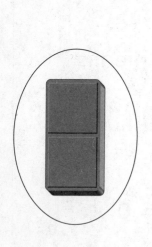

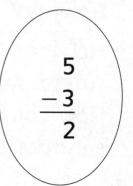

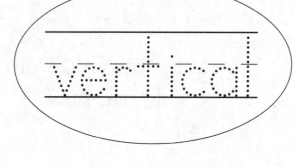

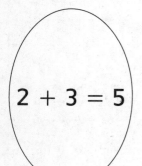

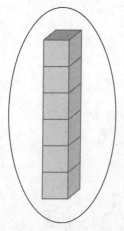

**Teacher Directions:** Provide a description, explanation, or example of the new term using images or real objects. Have students say the letters aloud as they trace the math term. Direct students to draw a line from each example of vertical to the word in the center. Then encourage students to describe their work to a peer.

NAME _____ DATE _____

# Lesson 6 Problem Solving
## STRATEGY: Draw a Diagram

<u>Underline</u> what you know. (Circle) what you need to find. Draw a diagram to solve.

1. There are **9** frogs on a tree.

    **4** of the frogs **hop away**.

    How many frogs **are left** on the tree?

 frog

 tree

There are ____ frogs **left** on the tree.

 **Teacher Directions:** Provide a description, explanation, or example of the boldface terms and nouns using images or real objects. Read each sentence and have students echo read. Encourage students to use the ten-frame to diagram the problem, and then write their answer in the restated question. Have students read the answer sentence aloud.

Grade 1 • Chapter 2 *Subtraction Concepts* 21

NAME _____ DATE _____

# Lesson 7 Vocabulary Definition Map
## Compare Groups

Use the definition map to write what the math word means and tell what the word is like. Write or draw a math example. Share your examples with a classmate.

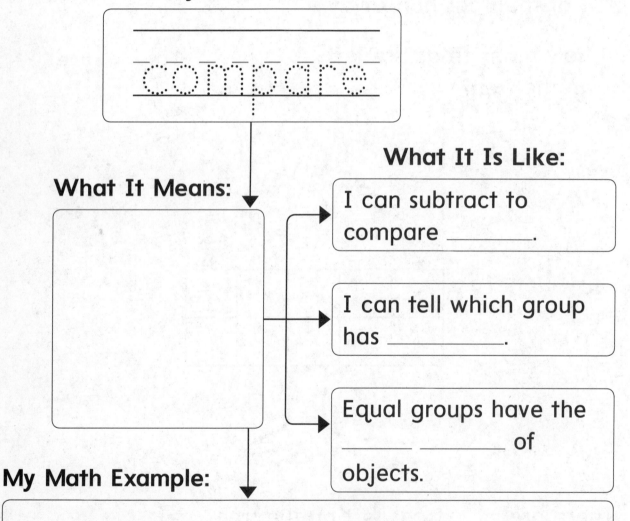

**My Math Word:** compare

**What It Means:**

**What It Is Like:**
- I can subtract to compare _____.
- I can tell which group has _____.
- Equal groups have the _____ _____ of objects.

**My Math Example:**

**Teacher Directions:** Provide a description, explanation, or example of the new term using images or real objects. Have students use the lesson or Glossary to define the math term. Direct students to list characteristics and draw a picture representing their math term. Then encourage students to describe their picture to a peer.

22 Grade 1 • Chapter 2 *Subtraction Concepts*

# Lesson 8 Difference Identification

*Subtract from 4 and 5*

**Draw a line to match to the difference.**

 =        3

 =        2

 =        0

 =        1

**Draw a picture to show one way to subtract from 5.**

**Teacher Directions:** Use manipulatives to show various ways to subtract from 4 and 5. Model a subtraction sentence for each and have students repeat it. Have students draw a line to match each subtraction sentence to the difference. Direct students to draw a picture representing a way to subtract from 5. Then encourage students to describe their picture to a peer.

Grade 1 • Chapter 2 *Subtraction Concepts*

# Lesson 9 Difference Identification
## Subtract from 6 and 7

Match a subtraction sentence to a way to subtract from 7.

     $7 - 5 = 2$

     $7 - 2 = 5$

     $7 - 3 = 4$

     $7 - 1 = 6$

Draw a picture to show one way to subtract from 6.

**Teacher Directions:** Use manipulatives to show various ways to subtract from 6 and 7. Model a subtraction sentence for each and have students repeat it. Have students draw a line to match each image to the correct subtraction sentence. Then direct students to draw a picture representing a way to subtract from 6. Finally encourage students to describe their picture to a peer.

# Lesson 10 Vocabulary Sentence Frames
## Subtract from 8

The math words in the word bank are for the sentences below. Write the words that fit in each sentence on the blank lines.

| Word Bank |
| :---: |
| minus      difference      equals |

1. The answer to a subtraction problem is the _____.
   8 − 5 = 3
       ↑

2. Eight minus four _____ four.

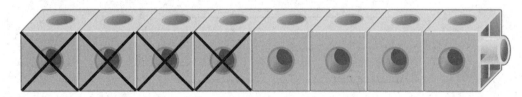

3. Eight _____ six equals two.

**Teacher Directions:** Provide a description, explanation, or example of the each term using images or real objects. Read each sentence frame and have students echo read. Direct students to write the correct terms in each blank. Then encourage students to read each sentence to a peer.

# Lesson 11 Word Web

*Subtract from 9*

Read the words in the word bank. Write the subtraction terms in the ovals.

| **Word Bank** |
|---|
| take away    plus    difference |
| are left    minus    sum    in all |

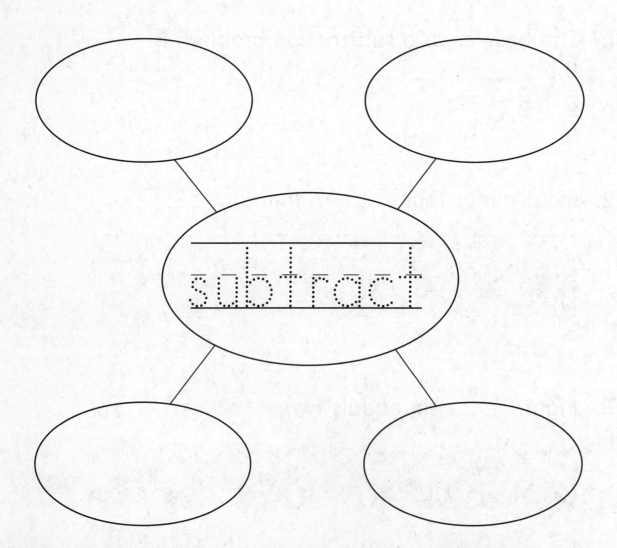

**Teacher Directions:** Provide a description, explanation, or example of the new term using images or real objects. Have students use the word bank to write subtraction terms in the ovals. Inform students that addition terms in the word bank will not be used. Then encourage students to describe the subtraction terms to a peer.

NAME _____ DATE _____

# Lesson 12 Vocabulary Word Study
*Subtract from 10*

Circle the correct word to complete the sentence.

1. The difference is the answer to a _____ problem.

      subtraction                addition

---

Show what you know about the word:

# difference

There are ____ letters.

There are ____ vowels.

There are ____ consonants.

____ letters − ____ vowels = ____ consonants.

---

Draw a picture to show what the word means.

**Teacher Directions:** Provide a description, explanation, or example of the new term using images or real objects. Read the sentence and have students circle the correct word. Direct students to count the letters, vowels and consonants in the math term, then complete the subtraction number sentence. Guide students to draw a picture representing their math term. Then encourage students to describe their picture to a peer.

Grade 1 • Chapter 2 *Subtraction Concepts*    27

NAME _____    DATE _____

# Lesson 13 Note Taking
## *Relate Addition to Subtraction*

Read the question. Write words you need help with. Use your lesson to write your Cornell notes. Write or draw math examples to explain your thinking. Share your examples with a classmate.

| **Building on the Essential Question** | **Notes:** |
|---|---|
| How can I relate addition to subtraction? | Related facts use the _____ _____. |
| | You can write related addition and subtraction _____. |
| | $2 + 7 = 9 \quad\quad 9 - 7 = 2$ |
| | $7 + 2 = 9 \quad\quad 9 - 2 = 7$ |
| **Words I need help with:** | These facts can help you _____ and _____. |
| | You can use $2 + 7 = 9$ to find $9 - 2 =$ _____. |

**My Math Examples:**

**Teacher Directions:** Read the Building on the Essential Question and have students list words/phrases they need assistance with. Provide descriptions, explanations, or examples of the terms using images or real objects. Read each sentence frame and have students write the appropriate terms. Have students read their notes aloud. Direct students to draw a picture representing the question. Then encourage students to describe their picture to a peer.

NAME _____ DATE _____

# Lesson 14 Note Taking
*True and False Statements*

Read the question. Write words you need help with. Use your lesson to write your Cornell notes. Write or draw math examples to explain your thinking.

| **Building on the Essential Question** How do I know if a statement is true or false? | **Notes:** Statements can be _____ or _____. A _____ statement is correct.     3 − 1 = 2 A _____ statement is **in**correct or wrong.     incorrect ⇩   3 − 1 = 1 I can use _____ _____ to see if the difference is correct. 1 + 2 = 3    3 − 1 = 2 2 + 1 = 3    3 − 2 = 1 |
|---|---|
| **Words I need help with:** | |

**My Math Examples:**

 **Teacher Directions:** Read the Building on the Essential Question and have students list words/phrases they need assistance with. Provide descriptions, explanations, or examples of the terms using images or real objects. Read each sentence frame and have students write the appropriate terms. Have students read their notes aloud. Direct students to draw a picture representing the question.

Grade 1 • Chapter 2 *Subtraction Concepts* 29

NAME _____  DATE _____

# Chapter 3 Addition Strategies to 20
*Inquiry of the Essential Question:*

## How do I use strategies to add numbers?

6¢      ___¢     ___¢     ___¢

I see ...

I think ...

I know ...

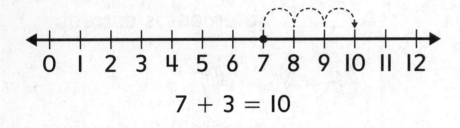

7 + 3 = 10

I see ...

I think ...

I know ...

8 + 8 = 16

8 + 9 = 17

8 + 7 = 15

I see ...

I think ...

I know ...

Questions I have...

_____
_____
_____
_____

**Teacher Directions:** Read the Essential Question for students. Have students echo read. Direct students to describe their observations, inferences, and prior knowledge of each math example. Encourage students to write or draw additional questions they may have. Then have students share their ideas/questions with a peer.

NAME _____ DATE _____

# Lesson 1 Note Taking

*Count On 1, 2, or 3*

Read the question. Write words you need help with. Use your lesson to write your Cornell notes. Write or draw math examples to explain your thinking.

| **Building on the Essential Question** How can I count on to add 1, 2, or 3? | **Notes:** I can count on to _____. <br> 7 + 2 = ? <br> 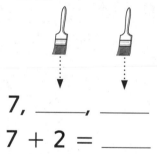 <br> I should start with the _____ number. <br> 7 + 🖌🖌 = ? <br> Then I should count on to add 2 _____. <br> 7, ____, ____ <br> 7 + 2 = ____ |
|---|---|
| **Words I need help with:** | |

**My Math Examples:**

**Teacher Directions:** Read the Building on the Essential Question and have students list words/phrases they need assistance with. Provide descriptions, explanations, or examples of the terms using images or real objects. Read each sentence frame and have students write the appropriate terms and numbers. Have students read their notes aloud. Direct students to draw a new picture representing the question. Then encourage students to describe their picture to a peer.

Grade 1 • Chapter 3 Addition Strategies to 20

NAME _____ DATE _____

# Lesson 2 Four-Square Vocabulary
*Count On Using Pennies*

Trace the word. Write the definition for *count on*. Write what the words mean, draw a picture, and write your own sentence using the words.

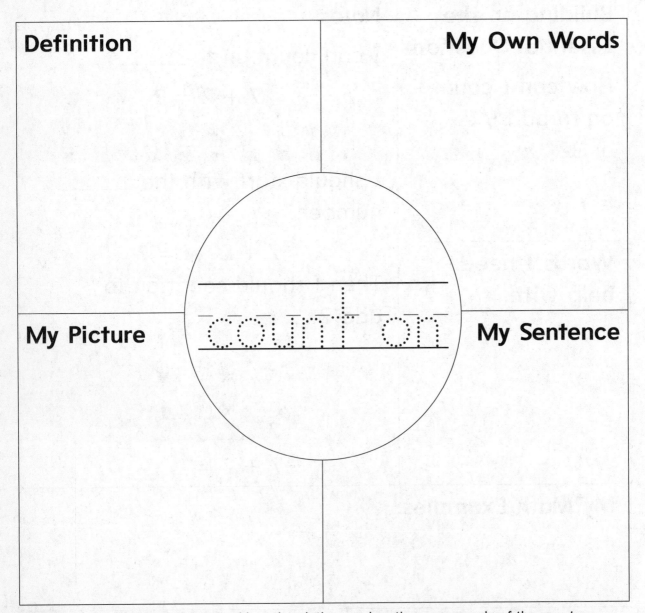

**Teacher Directions:** Provide a description, explanation, or example of the new term using images or real objects. Have students say the letters aloud as they trace the math term. Then have them use the Glossary to write the definition. Direct students to write a definition in their own words and draw a picture representing their math term. Have students write a sentence using the term and then encourage students to read their sentence to a peer.

NAME _____  DATE _____

# Lesson 3 Vocabulary Definition Map
*Use a Number Line to Add*

Trace the term. Use the definition map to write what the math term means and tell what the term is like. Write or draw a math example. Share your examples with a classmate.

**My Math Word:**

*number line* (traced)

**What It Means:**

**What It Is Like:**

I can use a number line to _____.

I should start with the _____ number.

Then I can _____ _____ by moving to the right.

**My Math Example:**

**Teacher Directions:** Provide a description, explanation, or example of the new term using images or real objects. Have students say the letters aloud as they trace the math term. Then have them use the lesson or Glossary to define it. Direct students to list characteristics and draw a picture representing their math term. Then encourage students to describe their picture to a peer.

Grade 1 • Chapter 3 Addition Strategies to 20

# Lesson 4 Vocabulary Sentence Frames
## Use Doubles to Add

The math words in the word bank are for the sentences below. Write the words that fit in each sentence on the blank lines.

| Word Bank | | |
|---|---|---|
| doubles | addends | sum |

1. The answer to an addition problem is the _____.

    4 + 5 = 9
        ↑

2. _____ are the numbers you add.

    4 + 5 = 9
    ↑   ↑

3. Both addends are the same in a _____ fact.

    (4 + 4) = 8

**Teacher Directions:** Provide a description, explanation, or example of the each term using images or real objects. Read each sentence frame and have students echo read. Direct students to write the correct terms in each blank. Then encourage students to read each sentence to a peer.

# Lesson 5 Vocabulary Identification
## Use Near Doubles to Add

Match each term to a picture.

doubles ● ● ● + ● ● ●

doubles plus 1 ● ● ● + ● ●

doubles minus 1

---

Write the correct term from above on the blank lines.

You can use near _____ facts to find a sum.

shows _____.

shows _____.

**Teacher Directions:** Review the terms using images or real objects. Have students say each term and then draw a line to match the term to its example. Model and prompt students to describe a near doubles fact in this way: *The first addend is _____. The next addend is _____. That is 1 [more/less]. So this is doubles [plus/minus] 1.* Have students write the correct term in each sentence. Encourage students to read the sentences to a peer.

NAME _____ DATE _____

# Lesson 6 Problem Solving
## STRATEGY: Act It Out

Underline what you know. Circle what you need to find. Act out the problem to find the answer.

1. A hot dog **vendor** sold **9** hot dogs on **Monday.**

   hot dog

   **She** (the vender) sold the **same number** of hot dogs on **Tuesday**.

   How many hot dogs did **she** sell **in all**?

   vendor

   Act it out

   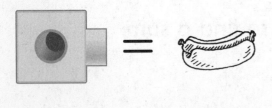

   ___ + ___ = ___

   She sold ___ hot dogs in all.

**Teacher Directions:** Provide a description, explanation, or example of the boldface terms and nouns using images or real objects. Read each sentence and have students echo read. Have students use connecting cubes to act out the problem. Then encourage them to use a part-part-whole mat to keep track of their addends and the sum. Direct them to write their answer in the restated question. Have students read the answer sentence aloud.

36 Grade 1 • Chapter 3 Addition Strategies to 20

# Lesson 7 Sum Identification
## Make 10 to Add

Match addition examples to show how to make a 10 to add.

[image of 8 cubes] + [image of 4 cubes] = 10 + 2

7 + 5 = 10 + 1

[image of 10 cubes] + [image of 4 cubes] = 10 + 4

9 + 5 = 10 + 3

Draw a picture to show one way to make a 10 to add.

**Teacher Directions:** Model an addition sentence, such as 8 + 3, and use manipulatives to show how to make a 10 to find the sum. Have students match an example on the left with the make-a-10 example on the right. Then direct students to draw a picture representing a way to make a 10 when adding. Finally encourage students to describe their picture to a peer.

Grade 1 • Chapter 3 Addition Strategies to 20

# Lesson 8 Concept Web
*Add in Any Order*

Use numbers or write a word from the word bank to complete each sentence in the concept web.

**Word Bank**

sum     addends     order

3 + ___ = 7
___ + 3 = 7

6 + ___ = 9
___ + 6 = 9

**Add in Any Order**

2 + 3 = 5

3 + 2 = 5

You can change the order of the ___ and get the same sum.

You can change the ___ of the addends and get the same sum.

You can change the order of the addends, and get the same ___.

**Teacher Directions:** Provide a description, explanation, or example of the terms in the word bank using images or real objects. Have students complete each addition sentence with the correct numbers. Then have them complete the remaining sentences using a word from the word bank. Finally encourage students to read the sentences to a peer.

38   Grade 1 • Chapter 3 Addition Strategies to 20

NAME _____ DATE _____

# Lesson 9 Note Taking
## *Add Three Numbers*

Read the question. Write words you need help with. Use your lesson to write your Cornell notes. Write or draw math examples to explain your thinking.

**Building on the Essential Question**

How can I use strategies to add three numbers?

**Words I need help with:**

**Notes:**

I can group _____.

I can add in any _____.

I can look for _____.

I can make a _____.

Then I can add the other number to find the _____.

**My Math Examples:**

**Teacher Directions:** Read the Building on the Essential Question and have students list words/phrases they need assistance with. Provide descriptions, explanations, or examples of the terms. Read each sentence frame and have students write the appropriate terms. Have students read their notes aloud. Direct students to draw a picture representing the question. Then encourage students to describe their picture to a peer.

Grade 1 • Chapter 3 Addition Strategies to 20    39

# Chapter 4 Subtraction Strategies to 20

*Inquiry of the Essential Question:*

**What strategies can I use to subtract?**

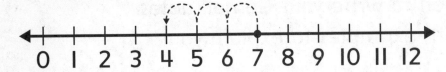

$7 - 3 = 4$

I see ...

I think ...

I know ...

$5 + 5 = 10$          $10 - 5 = 5$

I see ...

I think ...

I know ...

   $5 + 7 = 12$   $12 - 5 = 7$
         $7 + 5 = 12$   $12 - 7 = 5$

I see ...

I think ...

I know ...

Questions I have...

_____

_ _ _ _ _ _ _ _ _ _ _ _ _ _ _ _ _

_____

_____

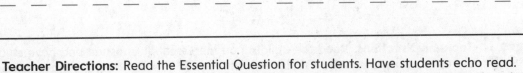

**Teacher Directions:** Read the Essential Question for students. Have students echo read. Direct students to describe their observations, inferences, and prior knowledge of each math example. Encourage students to write or draw additional questions they may have. Then have students share their ideas/questions with a peer.

# Lesson 1 Word Web

## Count Back 1, 2, or 3

Trace the math words. Draw a number story in each rectangle that shows the meaning of *count back*.

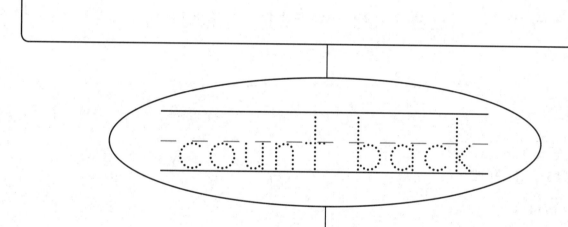

**Teacher Directions:** Provide a description, explanation, or example of the new term using images or real objects. Have students say the letters aloud as they trace the math term. Direct students to draw two picture stories that represent the math term. Then encourage students to tell their number stories to a peer.

Grade 1 • Chapter 4 Subtraction Strategies to 20   41

NAME _____ DATE _____

# Lesson 2 Note Taking

## Use a Number Line to Subtract

Read the question. Write words you need help with. Use your lesson to write your Cornell notes.

| **Building on the Essential Question** | **Notes:** |
|---|---|
| How can I use a number line to subtract? | **Word Bank**<br>count back   difference<br>greater number   subtract<br><br>$7 - 3 = $ _____<br><br>I can use a number line to help me _____.<br><br>I should start with the _____ _____.<br>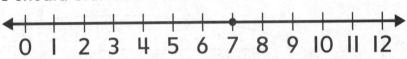<br>Then I can _____ _____ by moving to the left.<br><br>When I subtract, I can count back to find the _____.<br>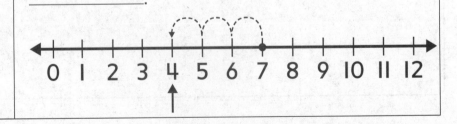 |
| **Words I need help with:** | |

 **Teacher Directions:** Read the Building on the Essential Question and have students list words/phrases they need assistance with. Provide descriptions, explanations, or examples of the terms using images or real objects. Read each sentence frame and have students write the appropriate terms. Have students read their notes aloud.

42  Grade 1 • Chapter 4 *Subtraction Strategies to 20*

NAME _____ DATE _____

# Lesson 3 Four-Square Vocabulary
## Use Doubles to Subtract

Trace the word. Write the definition for *doubles*. Write what the word means, draw a picture, and write your own sentence using the word.

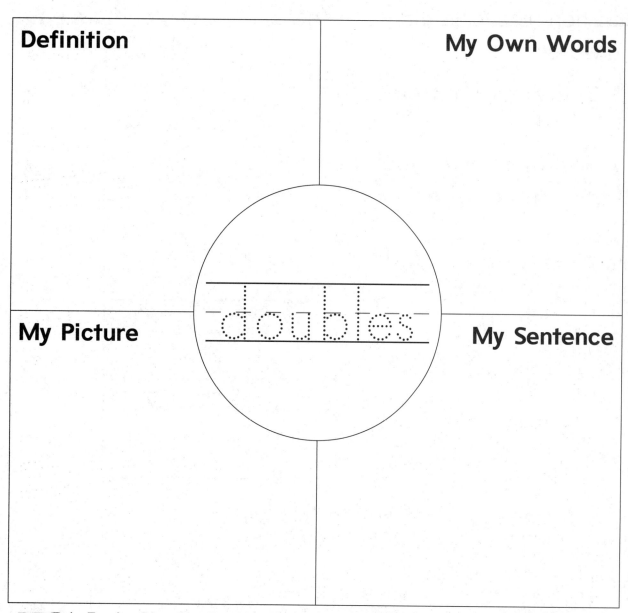

 **Teacher Directions:** Provide a description, explanation, or example of the new term using images or real objects. Have students say the letters aloud as they trace the math term. Then have them use the Glossary to write the definition. Direct students to write a definition in their own words and draw a picture representing their math term. Have students write a sentence using the term and then encourage students to read their sentence to a peer.

Grade 1 • Chapter 4 Subtraction Strategies to 20   43

# Lesson 4 Problem Solving

*STRATEGY: Write a Number Sentence*

<u>Underline</u> what you know. Ⓒircle what you need to find. Write a subtraction number sentence to solve.

1. There are **9** flamingos in the water.

   **5** flamingos **get out**.

   How many flamingos **are still in the water?**

 flamingo

 water

| Part | Part |
|------|------|
|      |      |
| Whole ||
|      |      |

____ ⊖ ____ ⊜ ____

____ flamingos **are still in the water**.

**Teacher Directions:** Provide a description, explanation, or example of the boldface terms and nouns using images or real objects. Read each sentence and have students echo read. Encourage students to use the part-part-whole mat and then write their answer in the restated question. Have students read the answer sentence aloud.

NAME _____ DATE _____

# Lesson 5 Difference Identification
## Make 10 to Subtract

Match addition examples to show how to make 10 to subtract.

13 − 5          = 10 − 1

13 − 6          = 10 − 2

15 − 6          = 10 − 5

12 − 7          = 10 − 3

Draw a picture to show one way to make 10 to subtract.

**Teacher Directions:** Model a subtraction sentence, such as 11 − 3, and use manipulatives to show how to make a 10 to find the difference. Have students match an example on the left with the make-a-10 example on the right. Direct students to draw a picture representing a way to make 10 when subtracting. Then encourage students to describe their picture to a peer.

Grade 1 • Chapter 4 Subtraction Strategies to 20   **45**

NAME _____ DATE _____

# Lesson 6 Note Taking

*Use Related Facts to Add and Subtract*

Read the question. Write words you need help with. Use your lesson to write your Cornell notes.

| **Building on the Essential Question** | **Notes:** |
|---|---|
| How can I use related facts to add and subtract? | **Word Bank**: addition   check   facts   related facts   subtraction <br><br> _____ _____ use the same numbers. <br><br> You can write related _____ and _____ facts. <br><br> $2 + 5 = 7$   $7 - 5 = 2$ <br> $5 + 2 = 7$   $7 - 2 = 5$ <br><br> These _____ can help you add and subtract. <br><br> You can use $2 + 5 = 7$ to find $7 - 2 = 5$. <br><br> You can use an addition fact to _____ your subtraction. |
| **Words I need help with:** | |

**Teacher Directions:** Read the Building on the Essential Question and have students list words/phrases they need assistance with. Provide descriptions, explanations, or examples of the terms using images or real objects. Read each sentence frame and have students write the appropriate terms. Have students read their notes aloud.

NAME _____  DATE _____

# Lesson 7 Concept Web

*Fact Families*

Trace the words in the center oval. Draw lines to match the true examples to the term *fact family*.

$3 + 4 = 7$
$4 + 3 = 7$
$7 + 3 = 10$
$10 - 3 = 7$

$4 + 5 = 9$
$5 + 4 = 9$
$9 - 4 = 5$
$9 - 5 = 4$

$6 + 2 = 8$
$5 + 3 = 8$
$4 + 4 = 8$
$7 + 1 = 8$

*fact family*

$6 + 5 = 11$
$5 + 6 = 11$
$11 - 6 = 5$
$11 - 5 = 6$

$1 + 5 = 6$
$5 + 1 = 6$
$6 - 5 = 1$
$6 - 1 = 5$

$7 + 8 = 15$
$8 + 7 = 15$
$15 - 7 = 8$
$15 - 8 = 7$

**Teacher Directions:** Provide a description, explanation, or example of the new term using images or real objects. Have students say the letters aloud as they trace the math term. Direct students to draw a line from each example of a fact family to the term in the center. Then encourage students to describe their work to a peer.

Grade 1 • Chapter 4 *Subtraction Strategies to 20*  **47**

# Lesson 8 Vocabulary Sentence Frames
*Missing Addends*

The math words in the word bank are for the sentences below. Write the words that fit in each sentence on the blank lines.

| Word Bank |
|---|
| count back    fact family    missing addend |

1. Addition and subtraction sentences that use the same numbers like

   $7 + 8 = 15$    $15 - 8 = 7$
   $8 + 7 = 15$    $15 - 7 = 8$

   are a _____ _____.

2. For $5 \oplus$ ____ $= 9$, the _____ _____ is 4.

3. On a number line, start at the greater number and _____ _____ if you want to subtract.

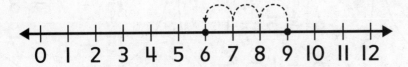

**Teacher Directions:** Provide a description, explanation, or example of the each term using images or real objects. Read each sentence frame and have students echo read. Direct students to write the correct terms in each blank. Then encourage students to read each sentence to a peer.

NAME _____  DATE _____

# Chapter 5 Place Value
*Inquiry of the Essential Question:*

**How can I use place value?**

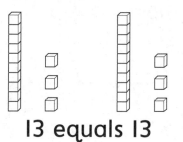

13 equals 13

I see ...

I think ...

I know ...

---

34 is greater than 21

I see ...

I think ...

I know ...

---

22 is less than 41

I see ...

I think ...

I know ...

---

Questions I have...

_ _ _ _ _ _ _ _ _ _ _ _ _ _ _ _ _ _ _ _ _ _ _ _ _ _ _ _ _ _

_ _ _ _ _ _ _ _ _ _ _ _ _ _ _ _ _ _ _ _ _ _ _ _ _ _ _ _ _ _

_ _ _ _ _ _ _ _ _ _ _ _ _ _ _ _ _ _ _ _ _ _ _ _ _ _ _ _ _ _

 **Teacher Directions:** Read the Essential Question for students. Have students echo read. Direct students to describe their observations, inferences, and prior knowledge of each math example. Encourage students to write or draw additional questions they may have. Then have students share their ideas/questions with a peer.

Grade 1 • Chapter 5 *Place Value*  49

NAME _____  DATE _____

# Lesson 1 Word Web
*Numbers 11 to 19*

Trace the math word. Draw a number story in each rectangle that shows the meaning of *more*.

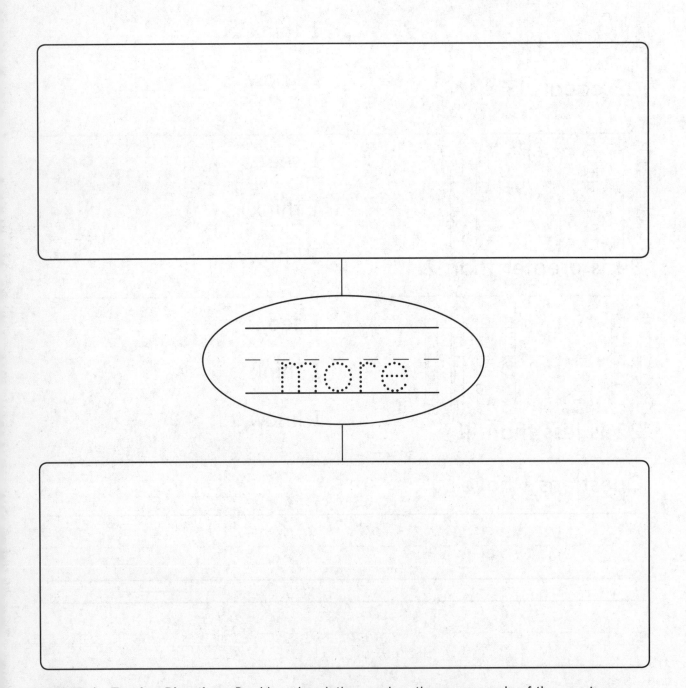

**Teacher Directions:** Provide a description, explanation, or example of the new term using images or real objects. Have students say the letters aloud as they trace the math term. Direct students to draw two picture stories that represent the math term. Then encourage students to tell their number stories to a peer.

50  Grade 1 • Chapter 5 *Place Value*

NAME _____ DATE _____

# Lesson 2 Number Identification
*Tens*

Trace each word. Then draw lines to match. The first one is done for you.

| Number | Tens | Word |
|---|---|---|
| 20 | 4 tens | forty |
| 40 | 5 tens | seventy |
| 70 | 2 tens | fifty |
| 50 | 8 tens | twenty |
| 80 | 7 tens | eighty |
| 60 | 1 ten | ten |
| 30 | 6 tens | ninety |
| 10 | 9 tens | thirty |
| 90 | 3 tens | sixty |

**Teacher Directions:** Use manipulatives to model each number. Model and then practice counting by tens as a group and then individually. First, have students trace each word, saying each letter as they write it. Have students say a number in the Number column, identify the matching number in the Tens column, and then draw a line to match the quantities. Students then match the Tens to the number word. Encourage partners to report about a number using a sentence frame such as: **Ninety is nine tens.**

Grade 1 • Chapter 5 *Place Value* **51**

NAME _____  DATE _____

# Lesson 3 Four-Square Vocabulary

*Count by Tens Using Dimes*

Trace the word. Write the definition for *dime*. Write what the word means, draw a picture, and write your own sentence using the word.

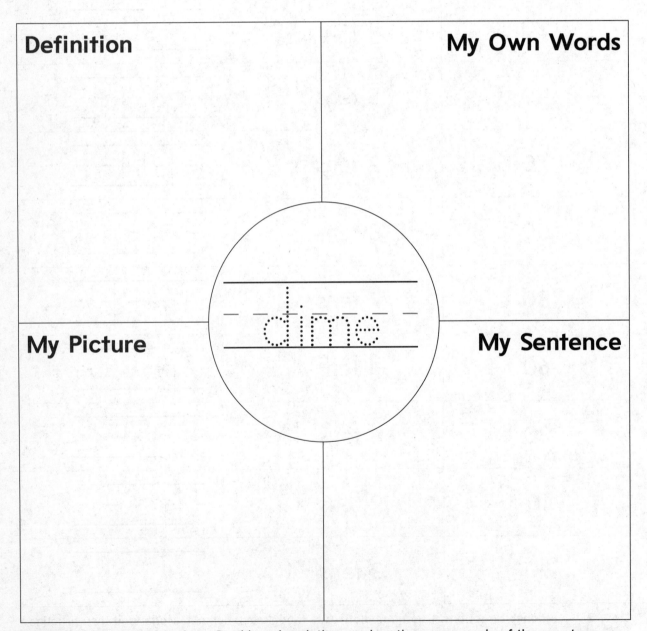

 **Teacher Directions:** Provide a description, explanation, or example of the new term using images or real objects. Have students use the Glossary to write the definition. Direct students to write a definition in their own words and draw a picture representing their math term. Have students write a sentence using the term and then encourage students to read their sentence to a peer.

NAME _____ DATE _____

# Lesson 4 Note Taking

*Tens and Some More*

Read the question. Write words you need help with. Use your lesson to write your Cornell notes. Write or draw math examples to explain your thinking.

**Building on the Essential Question**

How can I count ten and some more?

**Words I need help with:**

**Notes:**

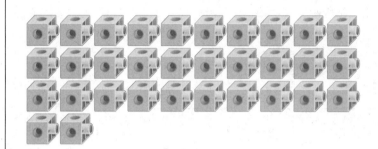

First, I circle groups of _____.

Then, I _____ the groups of ten.

Next, I count the _____.

Last, I write the _____.

____ tens and ____ ones is ____.

**My Math Examples:**

**Teacher Directions:** Read the Building on the Essential Question and have students list words/phrases they need assistance with. Provide descriptions, explanations, or examples of the terms using images or real objects. Read each sentence frame and have students write the appropriate terms. Have students read their notes aloud. Direct students to draw a picture representing the question. Then encourage students to describe their picture to a peer.

Grade 1 • Chapter 5 *Place Value* 53

NAME _____  DATE _____

## Lesson 5 Word Web
*Tens and Ones*

Use the word web to show examples of regrouping.

38 ones = _____ tens and 8 _____

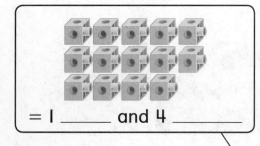
= 1 _____ and 4 _____

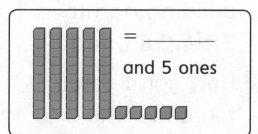
= _____ and 5 ones

regroup

74 ones = _____ tens and _____

_____ = 4 tens and 9 ones

2 _____ and _____ ones

**Teacher Directions:** Provide a description, explanation, or example of the new term using images or real objects. Have students complete the sentence frames to show examples of regrouping. Then encourage students to model regrouping strategies to a peer.

54 Grade 1 • Chapter 5 *Place Value*

# Lesson 6 Problem Solving
## STRATEGY: Make a Table

<u>Underline</u> what you know. (Circle) what you need to find. Make a table to solve.

1. **Mara** has 3 **groups** of 10 balls.

   **How many** balls does Mara have **in all**?

 ball

group of 10 balls

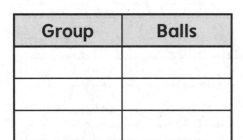

| Group | Balls |
|-------|-------|
|       |       |
|       |       |
|       |       |

Mara has ____ balls in all.

 **Teacher Directions:** Provide a description, explanation, or example of the boldface terms and nouns using images or real objects. Read each sentence and have students echo read. Encourage students to use the table to organize the information and then write their answer in the restated question. Have students read the answer sentence aloud.

# Lesson 7 Number Identification
## Numbers to 100

Trace each word. Then draw lines to match. The first one is done for you.

| Number | Tens and Ones | Word |
|---|---|---|
| 73 | 3 tens and 7 ones | seventy-three |
| 42 | 7 tens and 3 ones | thirty-seven |
| 37 | 4 tens and 2 ones | twenty-four |
| 24 | 2 tens and 4 ones | forty-two |
| 81 | 5 tens and 3 ones | eighty-one |
| 65 | 8 tens and 1 ones | fifty-three |
| 53 | 6 tens and 5 ones | sixteen |
| 16 | 9 tens and 8 ones | sixty-five |
| 98 | 1 ten and 6 ones | ninety-eight |

**Teacher Directions:** Use manipulatives to model each number. Model and then practice counting by tens as a group and then individually. First, have students trace each word, saying each letter as they write it. Have students say a number in the Number column, identify the matching number in the Tens column, and then draw a line to match the quantities. Students then match the Tens and Ones to the number word. Encourage partners to report about a number using a sentence frame such as: **37 is three tens and seven ones.**

NAME _____  DATE _____

# Lesson 8 Concept Web
*Ten More, Ten Less*

Trace the words. Then look at the number in each box. Write the number that is 10 more or 10 less.

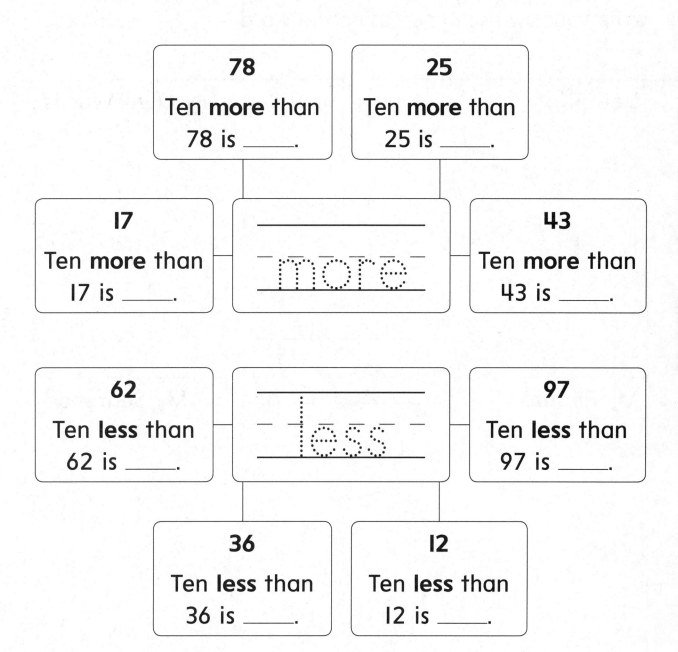

**Teacher Directions:** Provide math and non-math descriptions, explanations, or examples of the terms using images or real objects. Have students say each letter as they trace the term. Then have students complete the sentences in the web by writing a number that is ten more or ten less than the given number. Have students read their completed sentences to a peer.

Grade 1 • Chapter 5 *Place Value* 57

NAME _____  DATE _____

# Lesson 9 Four-Square Vocabulary
*Count by Fives Using Nickels*

Trace the word. Write the definition for *nickel*. Write what the word means, draw a picture, and write your own sentence using the word.

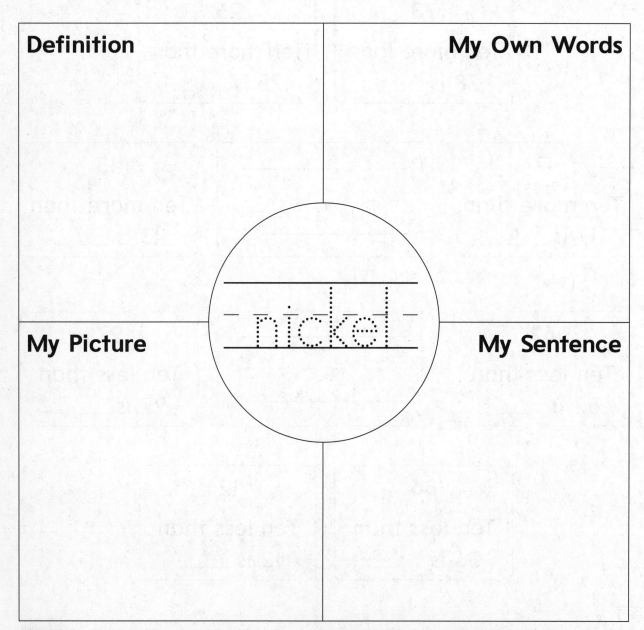

 **Teacher Directions:** Provide a description, explanation, or example of the new term using images or real objects. Have students use the Glossary to write the definition. Direct students to write a definition in their own words and draw a picture representing their math term. Have students write a sentence using the term and then encourage students to read their sentence to a peer.

NAME _____ DATE _____

# Lesson 10 Symbol Identification
## Use Models to Compare Numbers

Match each symbol to its meaning.

>           less than

<           equal to

=           greater than

---

Write the correct symbol meaning from above for each sentence on the blank lines.

1. 72 > 31     72 is _____ _____ 31.

2. 50 = 50     50 is _____ _____ 50.

3. 18 < 44     18 is _____ _____ 44.

**Teacher Directions:** Review the symbols to compare numbers using images or real objects. Have students say each symbol meaning then draw a line to match the symbol to its meaning. Direct students say each number comparison and then write the corresponding meanings in the sentences. Encourage students to read the sentences to a peer.

Grade 1 • Chapter 5 Place Value

# Lesson 11 Vocabulary Sentence Frames
*Use Symbols to Compare Numbers*

The math words and symbols in the word bank are for the sentences below. Write the words that fit in each sentence on the blank lines. Write the correct symbol between the pictures.

| **Word Bank** |
| :---: |
| equal to (=)    less than (<)    greater than (>) |

1. Seventy-five is _____ _____ fifty-seven.

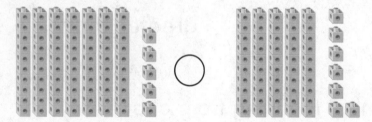

2. Thirty-three is _____ _____ thirty-three.

3. Nineteen is _____ _____ ninety.

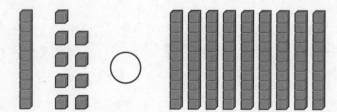

 **Teacher Directions:** Provide a description, explanation, or example of the each term using images or real objects. Read each sentence frame and have students echo read. Direct students to write the correct word bank terms and symbols in each blank. Then encourage students to read each sentence to a peer.

60  Grade 1 • Chapter 5 *Place Value*

# Lesson 12 Vocabulary Word Study
*Numbers to 120*

Circle the correct word to complete the sentence.

1. You can use hundreds, _____, and ones to show a three-digit number.

        tens        doubles        ones

---

Show what you know about the word:

## hundred

There are ____ letters.

There are ____ vowels.

There are ____ consonants.

____ vowels + ____ consonants = ____ letters in all.

---

Draw a picture to show what the word means.

**Teacher Directions:** Provide a description, explanation, or example of the new term using images or real objects. Read the sentence and have students circle the correct word. Direct students to count the letters, vowels and consonants in the math term, then complete the addition number sentence. Guide students to draw a picture representing their math term. Then encourage students to describe their picture to a peer.

# Lesson 13 Note Taking

*Count to 120*

Read the question. Write words you need help with. Use your lesson to write your Cornell notes.

| Building on the Essential Question | Notes: |
|---|---|
| How can I use strategies to count to 120? | **Word Bank**: chart, next, order |
| | I should always count in _____. |
| | 1, 2, 3, 4, 5    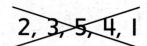 |
| **Words I need help with:** | I can use a number _____ if I need help. |
| | I say a number, and then I say the number that comes _____. |

NAME _____ DATE _____

# Lesson 14 Word Identification
*Read and Write Numbers to 120*

Match each word to its example.

before          7, ___, 9
                   ↑

after           7, ___, 9
                ↑

missing         7, ___, 9
                         ↑

---

Write the correct word from above for each sentence on the blank lines.

53, ___, 55

53 comes _____ the missing number.
55 comes _____ the missing number.
The _____ number is 54.

**Teacher Directions:** Review the words using images or real objects. Have students say each word and then draw a line to match the word to its example. Direct students say each word again and then complete the sentences. Encourage students to read the sentences to a peer.

NAME _____ DATE _____

# Chapter 6 Two-Digit Addition and Subtraction
*Inquiry of the Essential Question:*

**How can I add and subtract two-digit numbers?**

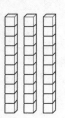 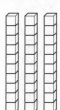

3 tens + 3 tens = 6 tens
30 + 30 = 60

I see …

I think …

I know …

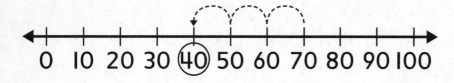

70 − 30 = 40

I see …

I think …

I know …

4 tens − 2 tens = 2 tens
40 − 20 = 20

I see …

I think …

I know …

Questions I have…

**Teacher Directions:** Read the Essential Question for students. Have students echo read. Direct students to describe their observations, inferences, and prior knowledge of each math example. Encourage students to write or draw additional questions they may have. Then have students share their ideas/questions with a peer.

64 Grade 1 • Chapter 6 *Two-Digit Addition and Subtraction*

NAME _____    DATE _____

# Lesson 1 Number and Word Identification
## Add Tens

Fill in the missing numbers and words. Then practice counting by tens.

| Number | Tens | Word |
|---|---|---|
| 20 | ____ tens | twenty |
| ____ | 3 tens | thirty |
| ____ | 4 tens | forty |
| 10 | ____ ten | ten |
| 50 | ____ tens | fifty |
| 70 | ____ tens | seventy |
| ____ | 9 tens | ninety |
| 80 | ____ tens | eighty |
| ____ | 6 tens | sixty |

**Teacher Directions:** Use manipulatives to model each number. Model and then practice counting by tens as a group and then individually. For each item, have students identify the number or word. Then have students fill in the missing number or word in the other two columns. Encourage partners to report back about a number using a sentence frame such as: [Ninety] is [nine] tens.

Grade 1 • Chapter 6 *Two-Digit Addition and Subtraction* 65

NAME _____  DATE _____

# Lesson 2 Concept Web
## Count On Tens and Ones

Use the word web to complete examples of counting on by tens and ones.

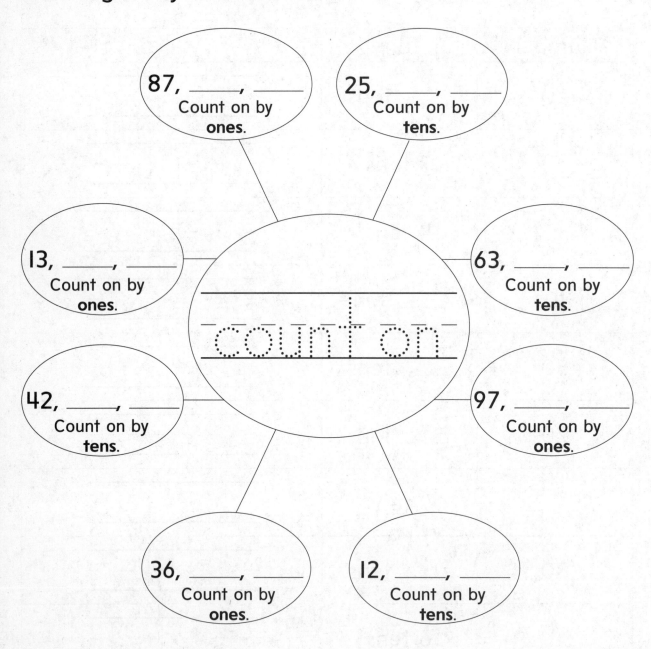

87, ____, ____ Count on by **ones**.

25, ____, ____ Count on by **tens**.

13, ____, ____ Count on by **ones**.

63, ____, ____ Count on by **tens**.

count on

42, ____, ____ Count on by **tens**.

97, ____, ____ Count on by **ones**.

36, ____, ____ Count on by **ones**.

12, ____, ____ Count on by **tens**.

**Teacher Directions:** Provide descriptions, explanations, or examples of the terms using images or real objects. Have students say each letter as they trace the term. Then have students fill in the numbers in the web by counting on by tens or ones. Have students check their answers with a peer.

66  Grade 1 • Chapter 6 Two-Digit Addition and Subtraction

NAME _____ DATE _____

# Lesson 3 Note Taking

*Add Tens and Ones*

Read the question. Write words you need help with. Use the lesson to write your Cornell notes.

| **Building on the Essential Question** How can I add tens and ones? | **Notes:** $32 + 5 = ?$<br>First, I should _____ each number.<br><table><tr><td>Tens</td><td>Ones</td></tr><tr><td>3</td><td>2</td></tr><tr><td>+</td><td>5</td></tr><tr><td></td><td></td></tr></table><br>Then, I should add the _____.<br><table><tr><td>Tens</td><td>Ones</td></tr><tr><td>3</td><td>2</td></tr><tr><td>+</td><td>5</td></tr><tr><td></td><td>7</td></tr></table><br>Next, I should add the _____.<br><table><tr><td>Tens</td><td>Ones</td></tr><tr><td>3</td><td>2</td></tr><tr><td>+</td><td>5</td></tr><tr><td>3</td><td>7</td></tr></table><br>The _____ is 3 tens and 7 ones. |
|---|---|
| **Words I need help with:** | |

**Teacher Directions:** Read the Building on the Essential Question and have students list words/phrases they need assistance with. Provide descriptions, explanations, or examples of the terms using images or real objects. Read each sentence frame and have students write the appropriate terms. Have students read their notes aloud.

Grade 1 • Chapter 6 *Two-Digit Addition and Subtraction* **67**

NAME _____ DATE _____

# Lesson 4 Problem Solving

STRATEGY: Guess, Check, and Revise

Underline what you know. Circle what you need to find. Solve the problem.

1. **Harper** draws **12** pictures.

   **Raven** draws **some** pictures.

   Together **they** (Harper and Raven) drew **15** pictures.

   How many pictures **did Raven draw**?

picture

| Part | Part |
|------|------|
|      |      |
| Whole ||
|       ||

___ ○ ___ ○ ___

Raven drew ___ pictures.

**Teacher Directions:** Provide a description, explanation, or example of the boldface terms and nouns using images or real objects. Read each sentence and have students echo read. Encourage students to use the part-part-whole mat to help them visualize what they know and need to know. Then have them write their answer in the restated question. Have students read the answer sentence aloud.

68 Grade 1 • Chapter 6 Two-Digit Addition and Subtraction

NAME _____ DATE _____

# Lesson 5 Vocabulary Definition Map
*Add Tens and Ones with Regrouping*

Use the definition map to write what the math word means and tell what the word is like. Write or draw a math example.

**My Math Word:**

regroup

**What It Means:**

**What It Is Like:**

*Re-* means "again." So *regroup* means "to _____ again."

When I regroup, I should _____ the ones.

Then I should regroup the 10 ones as I _____.

**My Math Example:**

**Teacher Directions:** Provide a description, explanation, or example of the term using images or real objects. Have students use the lesson or Glossary to define the math term. Direct students to list characteristics, and draw a picture representing their math term. Then encourage students to describe their picture to a peer.

Grade 1 • Chapter 6 Two-Digit Addition and Subtraction

NAME _____ DATE _____

# Lesson 6 Number Identification
*Subtract Tens*

Fill in the missing numbers. Then count back from 100 to 10.

| 1 | 2 | 3 | 4 | 5 | 6 | 7 | 8 | 9 | |
|---|---|---|---|---|---|---|---|---|---|
| 11 | 12 | 13 | 14 | 15 | 16 | 17 | 18 | 19 | |
| 21 | 22 | 23 | 24 | 25 | 26 | 27 | 28 | 29 | |
| 31 | 32 | 33 | 34 | 35 | 36 | 37 | 38 | 39 | |
| 41 | 42 | 43 | 44 | 45 | 46 | 47 | 48 | 49 | |
| 51 | 52 | 53 | 54 | 55 | 56 | 57 | 58 | 59 | |
| 61 | 62 | 63 | 64 | 65 | 66 | 67 | 68 | 69 | |
| 71 | 72 | 73 | 74 | 75 | 76 | 77 | 78 | 79 | |
| 81 | 82 | 83 | 84 | 85 | 86 | 87 | 88 | 89 | |
| 91 | 92 | 93 | 94 | 95 | 96 | 97 | 98 | 99 | 100 |

**Teacher Directions:** Use manipulatives to model each number. Model and then practice counting by tens as a group and then individually. Have students identify each missing number and write it in the blank. Count back from 100 to 10 by tens as a group. Encourage students to practice counting back by tens with a partner.

NAME _____ DATE _____

# Lesson 7 Note Taking

*Count Back by 10s*

Read the question. Write words you need help with. Use your lesson to write your Cornell notes.

| **Building on the Essential Question** How can I count back by 10s? | **Notes:** <br><br> **Word Bank** <br> number   difference   count back   subtract <br><br> $60 - 40 = ?$ <br><br> I can use a number line to _____ numbers by tens. <br><br> 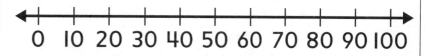 <br><br> I should start with the greater _____. <br><br> 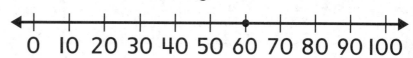 <br><br> Then I should _____ _____ by tens. <br><br> 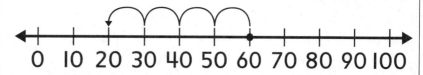 <br><br> Where I stop is the _____. <br> So, $60 - 40 = $ _____. |
|---|---|
| **Words I need help with:** | |

**Teacher Directions:** Read the Building on the Essential Question and have students list words/phrases they need assistance with. Provide descriptions, explanations, or examples of the terms using images or real objects. Read each sentence frame and have students write the appropriate terms. Have students read their notes aloud.

Grade 1 • Chapter 6 *Two-Digit Addition and Subtraction* 71

NAME _____ DATE _____

# Lesson 8 Four-Square Vocabulary
*Relate Addition and Subtraction of Tens*

Trace the words. Write the definition for *related facts*. Write what the word means, draw a picture, and write your own sentence using the word.

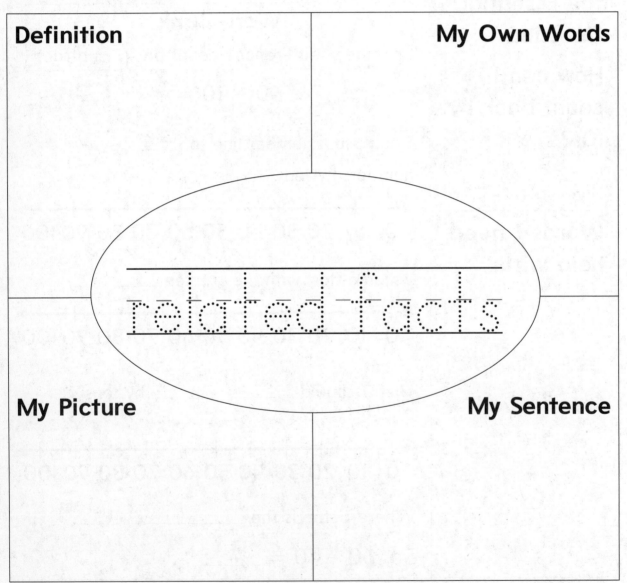

**Teacher Directions:** Provide a description, explanation, or example of the term using images or real objects. Have students use the Glossary to write the definition. Direct students to write a definition in their own words and draw a picture representing their math term. Have students write a sentence using the term and then encourage students to read their sentence to a peer.

NAME _____  DATE _____

# Chapter 7 Organize and Use Graphs
*Inquiry of the Essential Question:*

## How do I make and read graphs?

| Favorite Ice Cream | | |
|---|---|---|
| Ice Cream | Tally | Total |
| Chocolate | ||||| ||| | 8 |
| Vanilla | ||||| | 5 |
| Strawberry | ||| | 3 |

I see ...

I think ...

I know ...

**Favorite Season**

| | | | | | | | | | |
|---|---|---|---|---|---|---|---|---|---|
| Fall | 🍁 | 🍁 | 🍁 | | | | | | |
| Spring | 🌸 | 🌸 | 🌸 | 🌸 | 🌸 | 🌸 | | | |
| Summer | ☀ | ☀ | ☀ | ☀ | ☀ | ☀ | ☀ | | |

I see ...

I think ...

I know ...

**Favorite Ice Cream**

| Chocolate | | | | | | | | | |
|---|---|---|---|---|---|---|---|---|---|
| Vanilla | | | | | | | | | |
| Strawberry | | | | | | | | | |

0 1 2 3 4 5 6 7 8 9

I see ...

I think ...

I know ...

## Questions I have...

_____

_____

_____

**Teacher Directions:** Read the Essential Question for students. Have students echo read. Direct students to describe their observations, inferences, and prior knowledge of each math example. Encourage students to write or draw additional questions they may have. Then have students share their ideas/questions with a peer.

Grade 1 • Chapter 7 *Organize and Use Graphs* 73

# Lesson 1 Vocabulary Word Study
*Tally Charts*

Circle the correct word to complete the sentence.

1. A survey asks people the same ____.

   answer          question

Show what you know about the word:

## survey

There are ____ letters.

There are ____ vowels.

There are ____ consonants.

____ vowels + ____ consonants = ____ letters in all.

Draw a picture to show what the word means.

**Teacher Directions:** Provide a description, explanation, or example of the new term using images or real objects. Read the sentence and have students circle the correct word. Direct students to count the letters, vowels and consonants in the math term, then complete the addition number sentence. Guide students to draw a picture representing their math term. Then encourage students to describe their picture to a peer.

# Lesson 2 Problem Solving
*STRATEGY:* Make a Table

<u>Underline</u> what you know. (Circle) what you need to find. Make a table.

1. **Ana's** toy has 4 wheels.

   **Corey's** toy has 1 wheel.

   **Bryn's** toy has 2 wheels.

   The toys are a **unicycle**, a **bicycle**, and a **toy car**.

   Who has the bicycle?

unicycle

toy car

bicycle

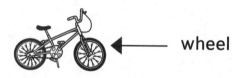

← wheel

| Name | Wheels | Riding Toy |
|---|---|---|
| Ana | | |
| Bryn | | |
| Corey | | |

_____ has the bicycle.

**Teacher Directions:** Provide a description, explanation, or example of the boldface terms and nouns using images or real objects. Read each sentence and have students echo read. Encourage students to use the table to organize the information and then write their answer in the restated question. Have students read the answer sentence aloud.

Grade 1 • Chapter 7 *Organize and Use Graphs* 75

# Lesson 3 Four-Square Vocabulary
## *Make Picture Graphs*

Trace the words. Write the definition for *picture graph*. Write what the words mean, draw a picture, and write your own sentence using the words.

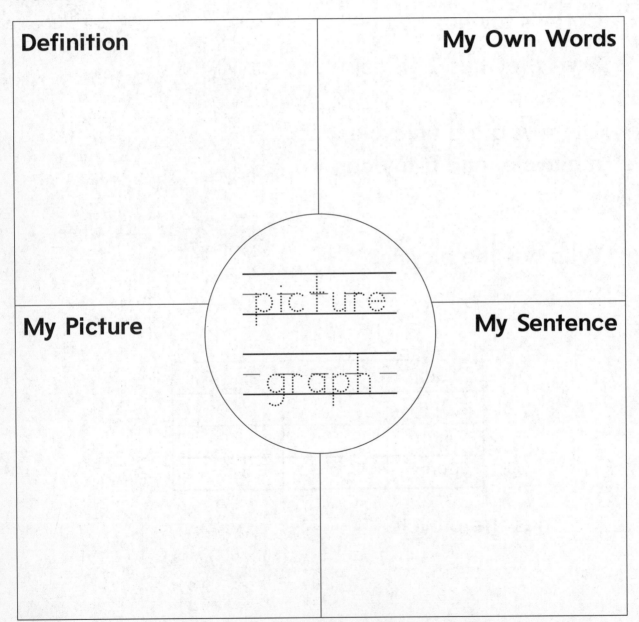

**Teacher Directions:** Provide a description, explanation, or example of the new term using images or real objects. Have students use the Glossary to write the definition. Direct students to write a definition in their own words and draw a picture representing their math term. Have students write a sentence using the term. Then encourage students to read their sentence to a peer.

NAME _____ DATE _____

# Lesson 4 Vocabulary Sentence Frames
*Read Picture Graphs*

The math words in the word bank are for the sentences below. Write the words that fit in each sentence on the blank lines.

| Word Bank | | |
|---|---|---|
| graph | data | picture graph |

1. A _____ _____ uses pictures to show information.

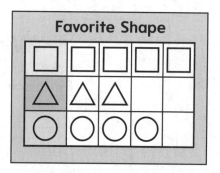

2. Information like: △△ is called _____.

3. A tally chart is a type of _____.

| Favorite Shape | | |
|---|---|---|
| Shape | Tally | Total |
| □ | IIII | 4 |
| △ | II | 2 |
| ○ | III | 3 |

**Teacher Directions:** Provide a description, explanation, or example of the each term using images or real objects. Read each sentence frame and have students echo read. Direct students to write the correct terms in each blank. Then encourage students to read each sentence to a peer.

Grade 1 • Chapter 7 *Organize and Use Graphs* 77

NAME _____ DATE _____

# Lesson 5 Vocabulary Definition Map
*Make Bar Graphs*

Use the definition map to write what the math word means and tell what the word is like. Write or draw a math example. Share your examples with a classmate.

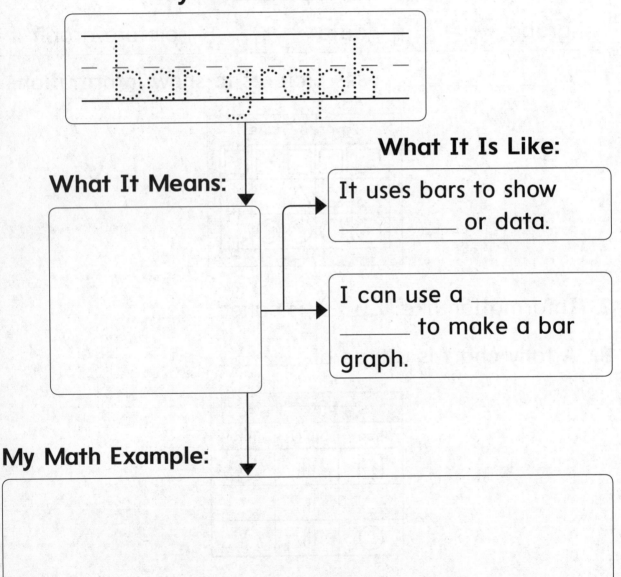

**My Math Word:** bar graph

**What It Means:**

**What It Is Like:**
It uses bars to show _____ or data.

I can use a _____ _____ to make a bar graph.

**My Math Example:**

**Teacher Directions:** Provide a description, explanation, or example of the new term using images or real objects. Have students use the lesson or Glossary to define the math term. Direct students to list characteristics, and draw a picture representing their math term. Then encourage students to describe their picture to a peer.

78  Grade 1 • Chapter 7 *Organize and Use Graphs*

NAME _____ DATE _____

# Lesson 6 Note Taking
## *Read Bar Graphs*

Read the question. Write words you need help with.
Use your lesson to write your Cornell notes.

| **Building on the Essential Question** | **Notes:** |
|---|---|
| How can I read bar graphs? | I know the bars on a bar graph tell ____ ____. |
| | The ____ on a bar graph can be horizontal or vertical. |
| | I should look where each bar ____. |
| **Words I need help with:** | Then I should read the ____. |
| | Favorite Fruit — horizontal bars — numbers  0 1 2 3 4 |

**Teacher Directions:** Read the Building on the Essential Question and have students list words/phrases they need assistance with. Provide descriptions, explanations, or examples of the terms using images or real objects. Read each sentence frame and have students write the appropriate terms. Have students read their notes aloud. Direct students to draw a picture representing the question. Then encourage students to describe their picture to a peer.

Grade 1 • Chapter 7 *Organize and Use Graphs*

NAME _____  DATE _____

# Chapter 8 Measurement and Time
*Inquiry of the Essential Question:*

**How do I determine length and time?**

I see ...

I think ...

I know ...

It's nine o'clock.   It's nine o'clock.

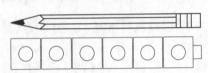

I see ...

I think ...

I know ...

The pencil is 6 connecting cubes long.

I see ...

I think ...

Longest                Shortest

I know ...

Questions I have...

_ _ _ _ _ _ _ _ _ _ _ _ _ _ _ _ _ _ _ _ _ _ _ _ _ _ _ _ _ _

_____

_ _ _ _ _ _ _ _ _ _ _ _ _ _ _ _ _ _ _ _ _ _ _ _ _ _ _ _ _ _

 **Teacher Directions:** Read the Essential Question for students. Have students echo read. Direct students to describe their observations, inferences, and prior knowledge of each math example. Encourage students to write or draw additional questions they may have. Then have students share their ideas/questions with a peer.

NAME _____  DATE _____

# Lesson 1 Word Identification
## Compare Lengths

Match.

short

long

length

---

Write the correct word from above for each sentence on the blank lines.

These cube trains are the **same** _____.

This cube train is _____.

This cube train is _____.

 **Teacher Directions:** Review the words using images or real objects. Have students say each word and then draw a line to match the word to its meaning. Direct students to say each word and then write the corresponding words in the sentence frames. Encourage students to read the sentences to a peer.

Grade 1 • Chapter 8 *Measurement and Time*  **81**

NAME _____ DATE _____

# Lesson 2 Concept Web
## Compare and Order Lengths

Look at the circled picture. Then circle the correct word.

 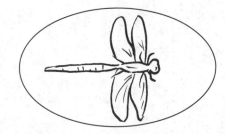 

long/longer/longest

 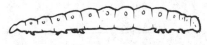 

short/shorter/shortest

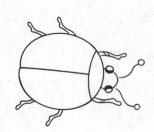 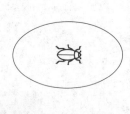 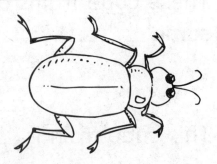

short/shorter/shortest

 **Teacher Directions:** Provide a description, explanation, or example of the new term using images or real objects. Direct students to look at the circled picture and then choose the correct word to describe it. Model and have students practice sentences that compare using sentence frames such as: **The butterfly is longer than the bee. The dragonfly is the longest.**

82 Grade 1 • Chapter 8 *Measurement and Time*

NAME _____ DATE _____

# Lesson 3 Vocabulary Definition Map
*Nonstandard Units of Length*

Use the definition map to write about the math word.

**My Math Word:**

measure

**What It Means:**

**What It Is Like:**

I measure to find the _____ of an object.

A paper clip or a penny is a nonstandard _____ I can use to measure.

I should _____ _____ the end of the object I want to measure with the unit I am measuring with.

**My Math Example:**

**Teacher Directions:** Provide a description, explanation, or example of the new term using images or real objects. Have students use the lesson or Glossary to define the math term. Direct students to list characteristics, and draw a picture representing their math term. Then encourage students to describe their picture to a peer.

Grade 1 • Chapter 8 *Measurement and Time* 83

# Lesson 4 Problem Solving

STRATEGY: *Guess, Check, and Revise*

Underline what you know. Circle what you need to find. Guess and measure.

1. **About** how many pennies long is the **object**?    penny

   Guess: about _____ pennies.

   Measure: about _____ pennies.

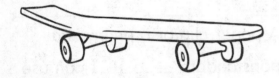

The **object** is **about** _____ pennies long.

# Lesson 5 Vocabulary Sentence Frames
*Time to the Hour: Analog*

The math words in the word bank are for the sentences below. Write the words that fit in each sentence on the blank lines.

| Word Bank | | |
|---|---|---|
| o'clock | minute hand | hour hand |

1. The long hand on the clock is the _____ _____.

2. It is two _____.

3. The hand that is pointing to 2 is the _____ _____.

**Teacher Directions:** Provide a description, explanation, or example of the each term using images or real objects. Read each sentence frame and have students echo read. Direct students to write the correct terms in each blank. Then encourage students to read each sentence to a peer.

Grade 1 • Chapter 8 *Measurement and Time*  85

NAME _____  DATE _____

# Lesson 6 Four-Square Vocabulary
*Time to the Hour: Digital*

Trace the words. Write the definition for *digital clock*. Write what the word means, draw a picture, and write your own sentence using the word.

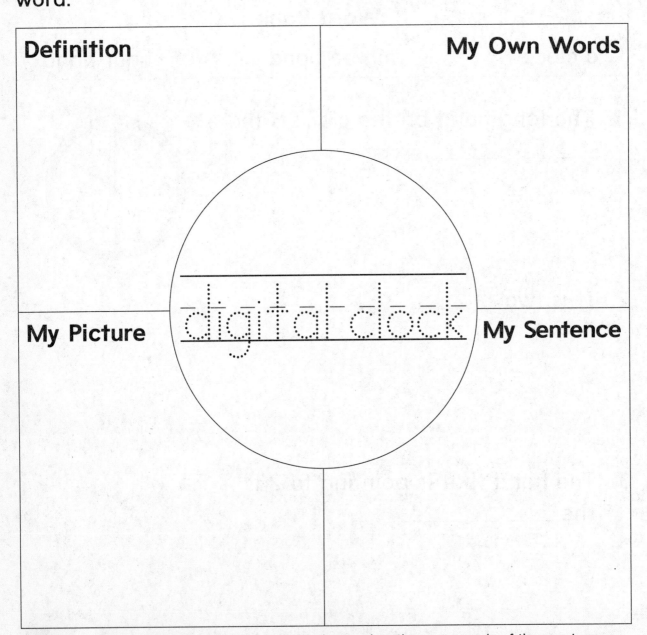

**Definition**

**My Own Words**

**My Picture**

**My Sentence**

**Teacher Directions:** Provide a description, explanation, or example of the new term using images or real objects. Have students use the Glossary to write the definition. Direct students to write a definition in their own words and draw a picture representing their math term. Have students write a sentence using the term. Then encourage students to read their sentence to a peer.

# Lesson 7 Concept Web
*Time to the Half Hour: Analog*

Trace the term. Draw a line from the term to each clock that shows a half hour.

**Teacher Directions:** Provide a description, explanation, or example of the new term using images or real objects. Direct students to trace the term in the box. Then have students draw a line from the term to each picture that shows time to the half hour. Model and have students practice sentences to describe the time, such as: **It is eight-thirty.** or **It is half past eight.**

Grade 1 • Chapter 8 *Measurement and Time*

NAME _____   DATE _____

# Lesson 8 Vocabulary Word Identification

*Time to the Half Hour: Digital*

Trace the words. Label the pictures with a term from the word bank.

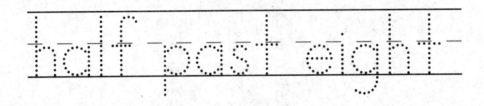

| Word Bank | | | |
|---|---|---|---|
| hour | hour hand | minutes | minute hand |

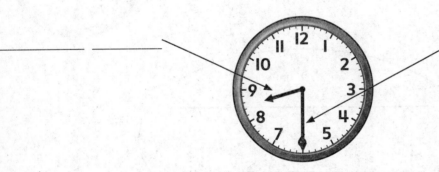

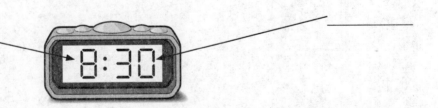

**Teacher Directions:** Review each term by providing a description, explanation, or example using images or real objects. Direct students to trace the math term *half past eight*. Then have students write a word from the word bank to label the parts of each picture. Then have students describe their work to a partner.

88  Grade 1 • Chapter 8 *Measurement and Time*

NAME _____  DATE _____

# Lesson 9 Note Taking

*Time to the Hour and Half Hour*

Read the question. Write words you need help with. Use your lesson to write your Cornell notes. Write or draw math examples to explain your thinking.

| **Building on the Essential Question**<br><br>How can I tell time to the hour and half hour? | **Notes:**<br><br>I know that one _____ is 60 minutes.<br><br>The _____ _____ points to the hour.<br><br>If the minute hand ends at 12, the clock shows an hour.<br><br>I know that a _____ _____ is 30 minutes.<br><br>If the minute hand ends at 6, it shows 30 _____. This is the half hour.<br><br>A digital clock shows the _____ on the left and the _____ on the right. |
|---|---|
| **Words I need help with:** | |
| **My Math Examples:** | |

**Teacher Directions:** Read the Building on the Essential Question and have students list words/phrases they need assistance with. Provide descriptions, explanations, or examples of the terms using images or real objects. Read each sentence frame and have students write the appropriate terms. Have students read their notes aloud. Direct students to draw a picture representing the question. Then encourage students to describe their picture to a peer.

Grade 1 • Chapter 8 *Measurement and Time*  89

NAME _____ DATE _____

# Chapter 9 Two-Dimensional Shapes and Equal Shares
*Inquiry of the Essential Question:*

How can I recognize two-dimensional shapes and equal shares?

3 Sides
3 Vertices

4 Sides
4 Vertices

I see ...

I think ...

I know ...

Rectangle      Rectangle

I see ...

I think ...

I know ...

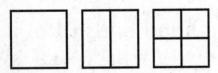

I see ...

I think ...

I know ...

Questions I have...

_____

- - - - - - - - - - - - - - - - - - - - - - - - -

_____

- - - - - - - - - - - - - - - - - - - - - - - - -

_____

**Teacher Directions:** Read the Essential Question for students. Have students echo read. Direct students to describe their observations, inferences, and prior knowledge of each math example. Encourage students to write or draw additional questions they may have. Then have students share their ideas/questions with a peer.

# Lesson 1 Word Web
## *Squares and Rectangles*

Label each item with a word from the word bank.

| Word Bank |
|---|
| rectangle     side     square     vertex |

(two-dimensional shapes web)

**Teacher Directions:** Provide a description, explanation, or example of the new terms using images or real objects. Teach students that the plural of vertex is vertices. Direct students to look at each picture and then label it with the correct word from the word bank. Model and have students practice sentences that describe each shape, such as: **A square has four sides and four vertices.**

Grade 1 • Chapter 9 *Two-Dimensional Shapes and Equal Shares*   **91**

NAME _____  DATE _____

# Lesson 2 Vocabulary Word Identification
*Triangles and Trapezoids*

Match each word to a picture.

triangle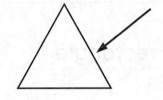

trapezoid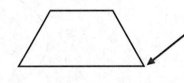

side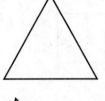

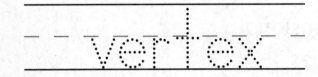vertex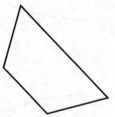

Write the correct word from the word bank for each sentence on the blank lines.

| Word Bank | | | |
|---|---|---|---|
| sides | triangle | trapezoid | vertices |

1. I have 4 _____ and 4 _____. I am a _____.

2. I have 3 _____ and 3 _____. I am a _____.

 **Teacher Directions:** Review the terms using images or real objects. Have students say each term then draw a line to match the term to a picture showing its meaning. Direct students say each term in the Word Bank and discuss plural forms of side (sides) and vertex (vertices). Have students write the corresponding terms in the sentences. Encourage students to read the sentences to a peer.

NAME _____ DATE _____

# Lesson 3 Vocabulary Definition Map
*Circles*

Use the definition map to write what the math word means and tell what the word is like. Write or draw a math example. Share your examples with a classmate.

**My Math Word:**

[circle]

**What It Means:**

**What It Is Like:**

A circle is _____ and _____.

A circle does **not** have _____.

A circle does **not** have _____.

**My Math Example:**

**Teacher Directions:** Provide a description, explanation, or example of the new term using images or real objects. Have students use the lesson or Glossary to define the math term. Direct students to list characteristics, and draw a picture representing their math term. Then encourage students to describe their picture to a peer.

Grade 1 • Chapter 9 *Two-Dimensional Shapes and Equal Shares* 93

NAME _____   DATE _____

# Lesson 4 Note Taking

## Compare Shapes

Read the question. Write words you need help with. Use your lesson to write your Cornell notes. Write or draw math examples to explain your thinking.

| **Building on the Essential Question** | **Notes:** |
|---|---|
| How can I compare two-dimensional shapes? | **Word Bank**<br>length    number    round<br>sides    straight    type<br><br>I can check if the shapes are _____ or if they have _____ sides.<br><br>I can count the number of _____.<br><br>I can check if the sides have the same _____.<br><br>I can count the _____ of vertices.<br><br>I can look for the same _____ of shape. |
| **Words I need help with:** | |

**My Math Examples:**

**Teacher Directions:** Read the Building on the Essential Question and have students list words/phrases they need assistance with. Provide descriptions, explanations, or examples of the terms using images or real objects. Read each sentence frame and have students write the appropriate terms. Have students read their notes aloud. Direct students to draw a picture representing the question. Then encourage students to describe their picture to a peer.

94    Grade 1 • Chapter 9 *Two-Dimensional Shapes and Equal Shares*

NAME _____ DATE _____

# Lesson 5 Vocabulary Word Study
*Composite Shapes*

Circle the correct word to complete the sentence.

1. I can _____ shapes to make a composite shape.

    put together              compare

---

Show what you know about the word:

## composite

There are _____ letters.

There are _____ vowels.

There are _____ consonants.

_____ vowels + _____ consonants = _____ letters in all.

---

Draw a picture to show what the word means.

**Teacher Directions:** Provide a description, explanation, or example of the new term using images or real objects. Read the sentence and have students circle the correct word. Direct students to count the letters, vowels and consonants in the math term, then complete the addition number sentence. Guide students to draw a picture representing their math term. Then encourage students to describe their picture to a peer.

Grade 1 • Chapter 9 *Two-Dimensional Shapes and Equal Shares*

NAME _____ DATE _____

# Lesson 6 Concept Web
## *More Composite Shapes*

Trace the word *shapes*. Circle the correct word.

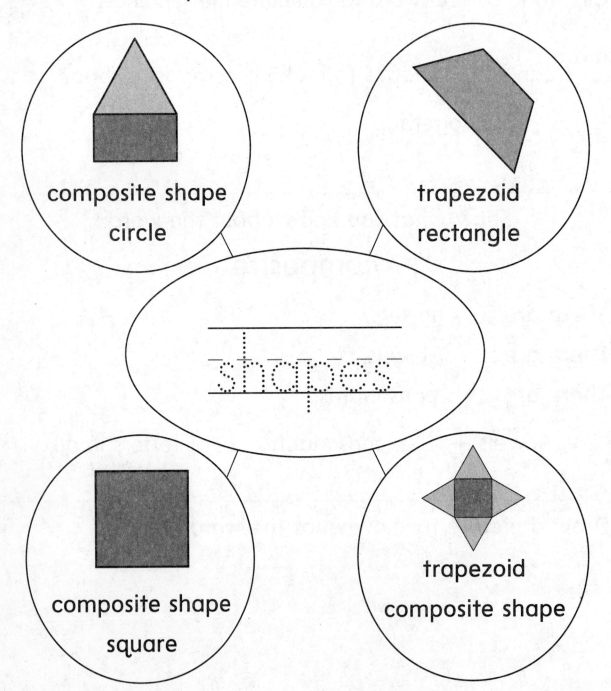

**Teacher Directions:** Provide a description, explanation, or example of the new terms using images or real objects. Direct students to look at each picture and then circle the correct word to describe it. Model and have students practice sentences that describe each shape, such as: **This composite figure is made of a square and four triangles.**

NAME _____  DATE _____

# Lesson 7 Problem Solving
*STRATEGY: Use Logical Reasoning*

<u>Underline</u> what you know. (Circle) what you need to find. Use logical reasoning to solve.

**1.** Rashad covered the pattern block with the **same three blocks.**

Circle which block he used.

triangle    square    trapezoid    rhombus

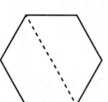 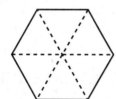 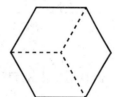 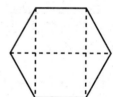

  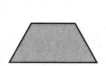 

Rashad used a _____.

 **Teacher Directions:** Provide a description, explanation, or example of the boldface terms and nouns using images or real objects. Read each sentence and have students echo read. Encourage students to use the patterns and then circle their answer. Model and teach sentence frames for describing students' work. For example, *I need two trapezoids to make the composite shape. That's not enough. I need six triangles to make the composite shape. That's too many.* Have partners practice describing the composite shapes.

Grade 1 • Chapter 9 *Two-Dimensional Shapes and Equal Shares* 97

# Lesson 8 Vocabulary Word Identification
## Equal Parts

Trace then match each word to the picture.

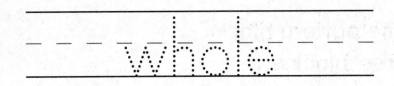 

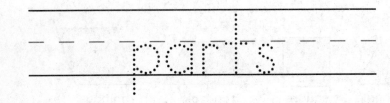 

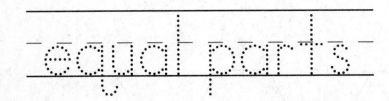 

Write the correct words from above for each sentence on the blank lines.

1. A whole can be separated into _____ _____.

2. Equal parts of the _____ are the same size.

3. Sometimes, _____ are not equal.

**Teacher Directions:** Review the new terms using images or real objects. Have students say each term and then draw a line to match each term to its meaning. Direct students to write the correct terms in the sentences. Encourage students to read the sentences to a peer.

98 Grade 1 • Chapter 9 *Two-Dimensional Shapes and Equal Shares*

NAME _____ DATE _____

# Lesson 9 Four-Square Vocabulary
*Halves*

Trace the word. Write the definition for *halves*. Write what the word means, draw a picture, and write your own sentence using the word.

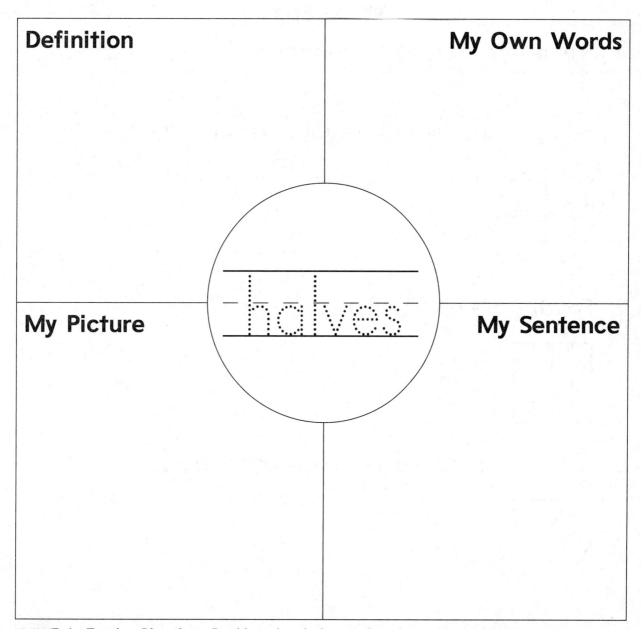

 **Teacher Directions:** Provide a description, explanation, or example of the new term using images or real objects. Have students use the Glossary to write the definition. Direct students to write a definition in their own words and draw a picture representing their math term. Have students write a sentence using the term. Then encourage students to read their sentence to a peer.

Grade 1 • Chapter 9 Two-Dimensional Shapes and Equal Shares

NAME _____ DATE _____

# Lesson 10 Vocabulary Sentence Frames
*Quarters and Fourths*

The math words in the word bank are for the sentences below. Write the words that fit in each sentence on the blank lines.

| **Word Bank** | | |
|---|---|---|
| quarters | fourths | halves |

1. _____ are the four equal parts of a whole.

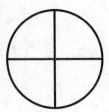

2. Fourths are also called _____.

3. _____ are two equal parts of a whole.

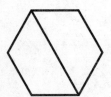

**Teacher Directions:** Provide a description, explanation, or example of the each term using images or real objects. Read each sentence frame and have students echo read. Direct students to write the correct terms in each blank. Then encourage students to read each sentence to a peer.

NAME _____   DATE _____

# Chapter 10 Three-Dimensional Shapes
*Inquiry of the Essential Question:*

## How can I identify three-dimensional shapes?

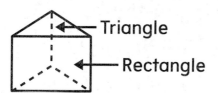

I see ...

I think ...

I know ...

---

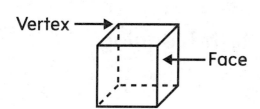

I see ...

I think ...

I know ...

---

Cone   Cone

I see ...

I think ...

I know ...

Questions I have...

_____
_____
_____
_____

**Teacher Directions:** Read the Essential Question for students. Have students echo read. Direct students to describe their observations, inferences, and prior knowledge of each math example. Encourage students to write or draw additional questions they may have. Then have students share their ideas/questions with a peer.

Grade 1 • Chapter 10 *Three-Dimensional Shapes* 101

NAME _____ DATE _____

# Lesson 1 Multiple Meaning Word
## Cubes and Prisms

Trace the word. Say the math word. Draw a picture that shows the math word meaning in the first box. Then draw a picture that shows a non-math word meaning in the other box.

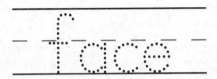

| Math Meaning | Non-Math Meaning |
|---|---|
|  |  |

Use the sentence frame below to help you describe your pictures.

This picture for the word _____ shows _____.

**Teacher Directions:** Provide math and non-math descriptions, explanations, or examples of the new term using images or real objects. Have students say then write the term. Then direct students to draw pictures showing a math and non-math meaning of the math term. Encourage them to describe their pictures to a peer using the sentence frame.

102 Grade 1 • Chapter 10 Three-Dimensional Shapes

## Lesson 2 Vocabulary Word Identification
*Cones and Cylinders*

Match.

  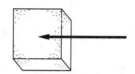

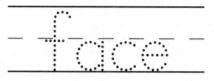  

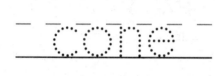  

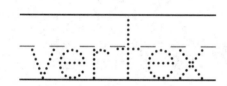  

Write the correct word from the word bank for each sentence on the blank lines.

| Word Bank | | | |
|---|---|---|---|
| cone | cylinder | face/faces | vertex/vertices |

1. I have 2 _____ and 0 _____. I am a _____.

2. I have 1 _____ and 1 _____. I am a _____.

**Teacher Directions:** Review the shapes using images or real objects. Have students say each shape name then draw a line to match the word to its corresponding image. Direct students say each term in the word bank, then write the corresponding terms in the sentences. Encourage students to read the sentences to a peer.

NAME _____ DATE _____

# Lesson 3 Problem Solving

## STRATEGY: Look for a Pattern

Underline what you know. Circle what you need to find. Find a pattern to solve.

1. Jenica made this pattern.

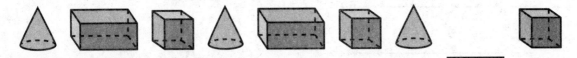

What shape is missing? Circle it.

cone          rectangular prism          cube

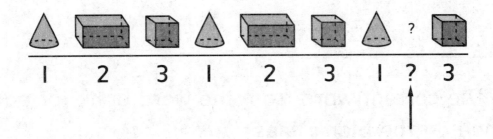

The _____ is missing.

**Teacher Directions:** Provide a description, explanation, or example of the bold face terms and nouns using images or real objects. Read each sentence and have students echo read. Encourage students to look for a pattern, discover the missing shape, circle the correct answer, and then write the shape name in the answer sentence. Have partners practice describing each shape and the pattern.

NAME _____ DATE _____

# Lesson 4 Note Taking
## Combine Three-Dimensional Shapes

Read the question. Write words you need help with. Use your lesson to write your Cornell notes. Write or draw math examples to explain your thinking.

| **Building on the Essential Question** | **Notes:** |
|---|---|
| How can I combine three-dimensional shapes? | I can _____ _____ three-dimensional shapes to make composite shapes.  |
| | I can _____ shapes on top of each other.  |
| **Words I need help with:** | It would be difficult to stack a cube on top of a _____.  |
| **My Math Examples:** | |

 **Teacher Directions:** Read the Building on the Essential Question and have students list words/phrases they need assistance with. Provide descriptions, explanations, or examples of the terms using images or real objects. Read each sentence frame and have students write the appropriate terms. Have students read their notes aloud. Direct students to draw a picture representing the question. Then encourage students to describe their picture to a peer.

Grade 1 • Chapter 10 *Three-Dimensional Shapes*

# What are VKVs® and How Do I Create Them?

Visual Kinethestic Vocabulary Cards® are flashcards that animate words by focusing on their structure, use, and meaning. The VKVs in this book are used to show cognates, or words that are similar in Spanish and English.

## Step 1
Go to the back of your book to find the VKVs for the chapter vocabulary you are currently studying. Follow the cutting and folding instructions at the top of the page. The vocabulary word on the BLUE background is written in English. The Spanish word is on the ORANGE background.

## Step 2
There are exercises for you to complete on the VKVs. When you understand the concept, you can complete each exercise. All exercises are written in English and Spanish. You only need to give the answer once.

## Step 3
Individualize your VKV by writing notes, sketching diagrams, recording examples, and forming plurals.

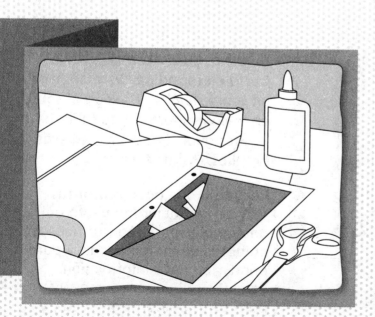

### How Do I Store My VKVs?
Take a 6" x 9" envelope and cut away a V on one side only. Glue the envelope into the back cover of your book. Your VKVs can be stored in this pocket!

Remember you can use your VKVs ANY time in the school year to review new words in math, and add new information you learn. Why not create your own VKVs for other words you see and share them with others!

Visual Kinesthetic Learning **VKVI**

# ¿Qué son las VKV y cómo se crean?

Las tarjetas de vocabulario visual y cinético (VKV) contienen palabras con animación que está basada en la estructura, uso y significado de las palabras. Las tarjetas de este libro sirven para mostrar cognados, que son palabras similares en español y en inglés.

### Paso 1
Busca al final del libro las VKV que tienen el vocabulario del capítulo que estás estudiando. Sigue las instrucciones de cortar y doblar que se muestran al principio. La palabra de vocabulario con fondo AZUL está en inglés. La de español tiene fondo NARANJA.

### Paso 2
Hay ejercicios para que completes con las VKV. Cuando entiendas el concepto, puedes completar cada ejercicio. Todos los ejercicios están escritos en inglés y español. Solo tienes que dar la respuesta una vez.

### Paso 3
Da tu toque personal a las VKV escribiendo notas, haciendo diagramas, grabando ejemplos y formando plurales.

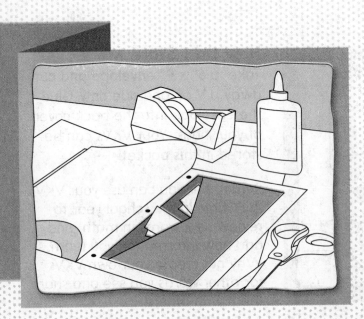

### ¿Cómo guardo mis VKV?
Corta en forma de "V" el lado de un sobre de 6" X 9". Pega el sobre en la contraportada de tu libro. Puedes guardar tus VKV en esos bolsillos. ¡Así de fácil!

Recuerda que puedes usar tus VKV en cualquier momento del año escolar para repasar nuevas palabras de matemáticas, y para añadir la nueva información. También puedes crear más VKV para otras palabras que veas, y poder compartirlas con los demás.

Chapter 1

✂ cut on all dashed lines   📄 fold on all solid lines

**más**

**+**

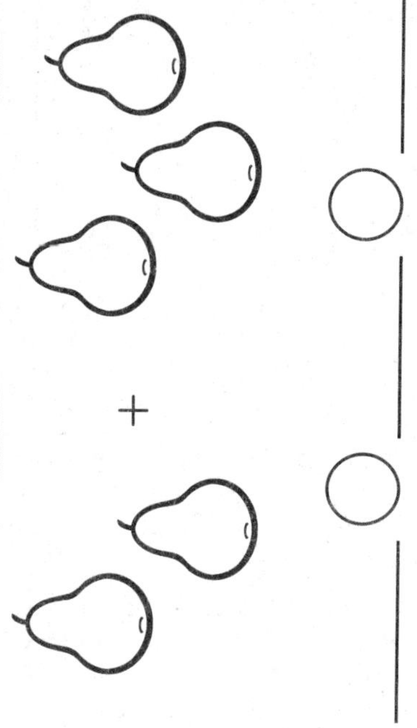

To find the whole, you add the (Para hallar el entero, sumas las) _____

**add**

**parte**

**part**

Chapter 1 Visual Kinesthetic Learning   VKV3

Chapter 1

cut on all dashed lines    fold on all solid lines

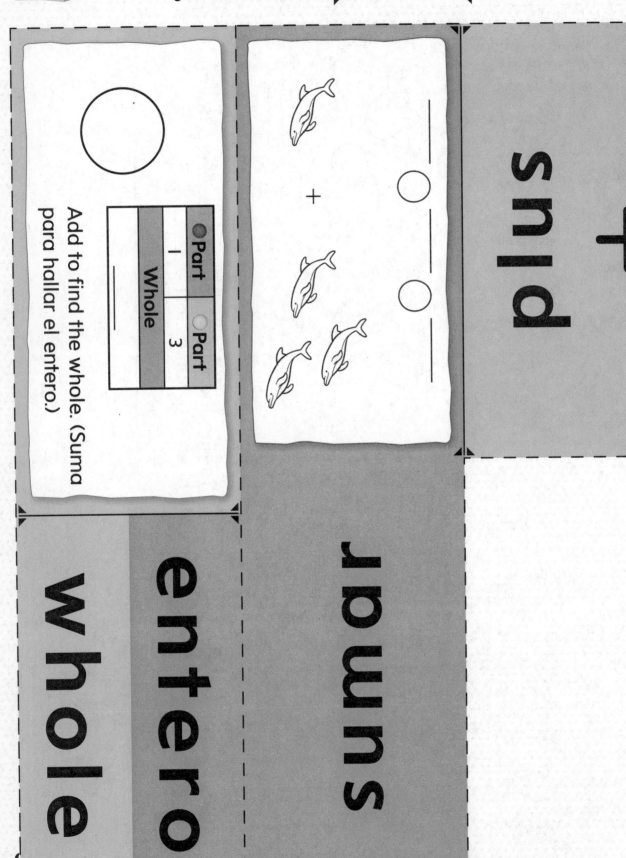

Add to find the whole. (Suma para hallar el entero.)

| Part | 1 | Whole |
|---|---|---|
| Part | 3 | |

plus +

sumar

entero

whole

Chapter 1

cut on all dashed lines   fold on all solid lines

**True or false? (¿Verdadero o falso?)**

$2 + 6 = 7$    false (falso)

$4 + 1 = 5$    true (verdadero)    false (falso)

$1 + 8 = 9$    true (verdadero)    false (falso)

$3 + 4 = 6$    true (verdadero)    false (falso)

**false**

**true**

**zero**

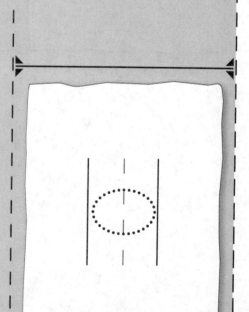

Chapter 1 Visual Kinesthetic Learning   VKV5

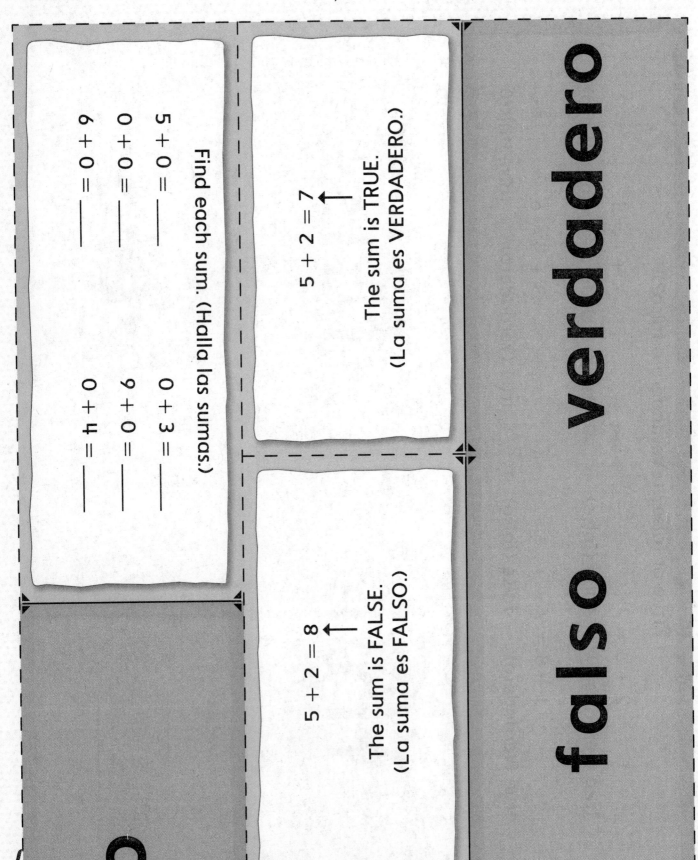

Chapter 2

cut on all dashed lines   fold on all solid lines

# difference

# minus —

Circle the difference. (Encierra en un círculo la diferencia.)

8 − 5 = 3

Write the subtraction number sentence. (Escribe el enunciado de resta.)

Chapter 2 Visual Kinesthetic Learning   VKV7

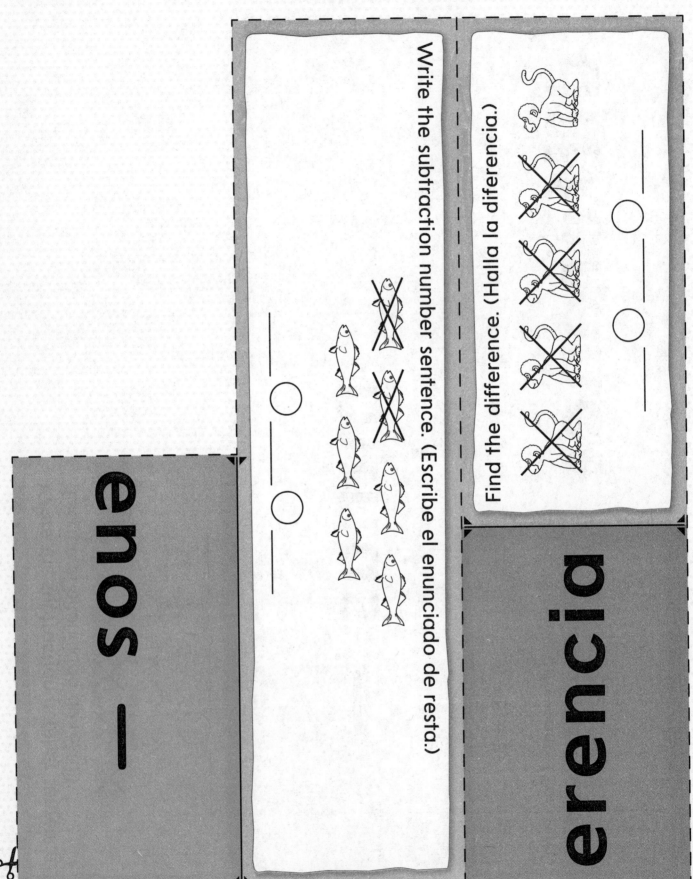

Chapter 2

related facts

number sentence

**Write both facts.** (Escribe ambas operaciones.)

| ● Part | 3 |
|---|---|
| Whole | 5 |
| ● Part | 2 |

___ + ___ = ___

___ − ___ = ___

**Write a number sentence.** (Escribe un enunciado numérico.)

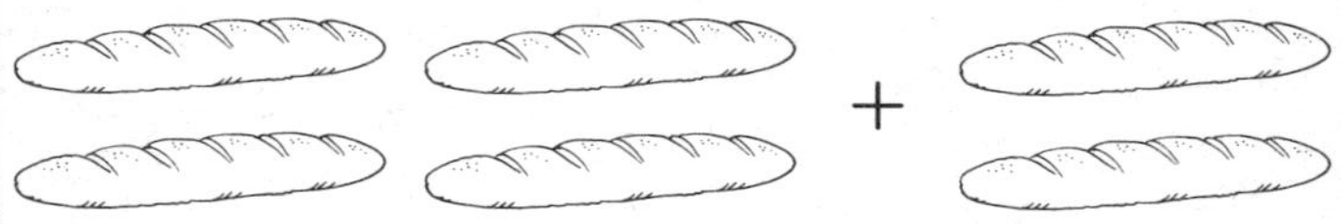

○ ___ + ○ ___ = ___

Chapter 2

✂ cut on all dashed lines    fold on all solid lines

**operaciones relacionadas**

Write the related subtraction facts. (Escribe las operaciones de resta relacionadas.)

$5 + 2 = 7$

___ − ___ = ___

___ − ___ = ___

Write a number sentence. (Escribe un enunciado numérico.)

**enunciado numérico**

VKV10  Chapter 2 Visual Kinesthetic Learning

Chapter 3

✂ cut on all dashed lines    fold on all solid lines

1 más uno

1

doubles minus one

Write each doubles fact. (Escribe las sumas de dobles.)

___ + ___ = 4     ___ + ___ = 8
___ + ___ = 12    ___ + ___ = 6
___ + ___ = 14    ___ + ___ = 10

Chapter 3 Visual Kinesthetic Learning   VKVII

**Chapter 3**

cut on all dashed lines    fold on all solid lines

1 plus one

1 dobles menos uno

Use doubles facts to add. (Usa suma de dobles para sumar.)

3 + 2 = ___     3 + 4 = ___
6 + 7 = ___     4 + 5 = ___
7 + 8 = ___     6 + 5 = ___

VKV12   Chapter 3 Visual Kinesthetic Learning

Chapter 4

cut on all dashed lines  fold on all solid lines

Count back to subtract. (Cuenta hacia atrás para restar.)

10, ____, ____, 5, ____
10 − 2 = ____
8, ____, ____, 9, ____
8 − 3 = ____
9 − 1 = ____

count back

missing addend

sumando desconocido

cut on all dashed lines

fold on all solid lines

Find the missing addend. (Halla el sumando desconocido.)

| Part | Part |
|---|---|
| | 6 |
| Whole | |
| 14 | |

___ + 6 = 14

14 − ___ = 6

$2 + 3 = 5$

# contar hacia atrás

Amelia had 7 books. She gave 3 books to Max. How many books does she have left? (Amelia tenía 7 libros. Le dio 3 libros a Max. ¿Cuántos libros le quedan?)

7, ___, ___, ___

Amelia has ___ books left. (A Amelia le quedan ___ libros.)

Chapter 5

cut on all dashed lines    fold on all solid lines

Circle groups of ten. Write how many tens and ones. (Encierra en un círculo los grupos de diez. Escribe cuántas decenas y cuántas unidades hay.)

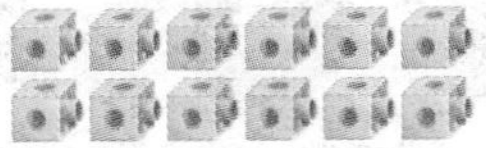

_____ ten (decenas)
_____ ones (unidades)

3 is less than 5.
(3 es menor que 5.)

regroup

mayor

less than

>

<

Chapter 5 Visual Kinesthetic Learning  VKV15

Chapter 5

cut on all dashed lines    fold on all solid lines

5 is greater than 3.
(5 es mayor que 3.)

agrupar

menor que

greater

<

>

Write how many tens and ones. (Escribe cuántas decenas y cuántas unidades hay.)

34 ones (unidades) = ___ tens and (decenas y) ___ ones (unidades) = ___

27 ones (unidades) = ___ tens and (decenas y) ___ ones (unidades) = ___

## Chapter 7

**cut on all dashed lines**  **fold on all solid lines**

---

**data**

Which is another word for *data*? (¿Cuál es otra palabra para decir *datos*?)

graph (gráfica)

information (información)

tally chart (tabla de conteo)

---

**tally chart**

How many people were surveyed? (¿Cuántas personas fueron encuestadas?) _____

| Favorite Shape | | |
|---|---|---|
| Shape | Tally | Total |
| △ Triangle | ||||| |
| ● Circle | | | |
| ■ Square | ||| | |

---

**survey**

A survey asks people the same (Un encuesta hace a las personas la misma) _____.

The votes of a survey can be marked in a (Los votos de una encuesta pueden marcarse en una) _____.

Chapter 7 Visual Kinesthetic Learning  VKV17

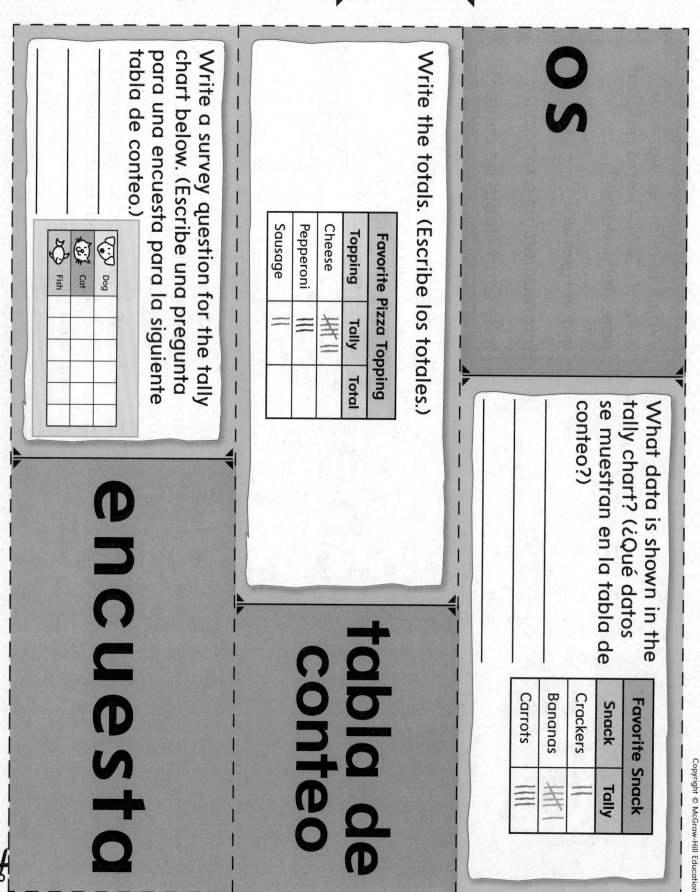

Chapter 7

✂ cut on all dashed lines   📄 fold on all solid lines

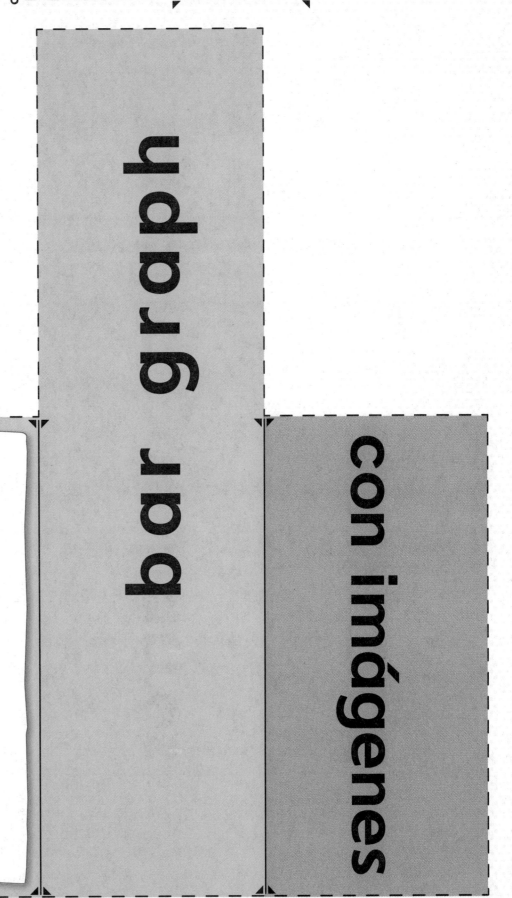

bar graph

con imágenes

Which snack was the favorite? (¿Cuál merienda era la favorita?)

Chapter 7 Visual Kinesthetic Learning  VKV19

Chapter 7

cut on all dashed lines    fold on all solid lines

gráfica de barras

picture

A picture graph uses _____ to show data. (Una gráfica con imágenes usa _____ para mostrar datos.)

A bar graph uses _____ to show data. (Una gráfica de barras usa _____ para mostrar datos.)

VKV20  Chapter 7 Visual Kinesthetic Learning

Chapter 8

cut on all dashed lines    fold on all solid lines

About how many cubes long is the nail? (¿Aproximadamente cuántos cubos de largo mide el clavo?)

_____ cubes (cubos)

Talia used a paperclip to measure a pencil. The pencil was about 4 paperclips long. Complete the equation. (Talia usó un sujetapapeles para medir un lápiz. El lápiz medía unos 4 sujetapapeles de largo. Completa la ecuación.)

1 unit (1 unidad) = _____

half

Write the time on the digital clock. (Escribe la hora en el reloj digital.)

measure

unit

hora

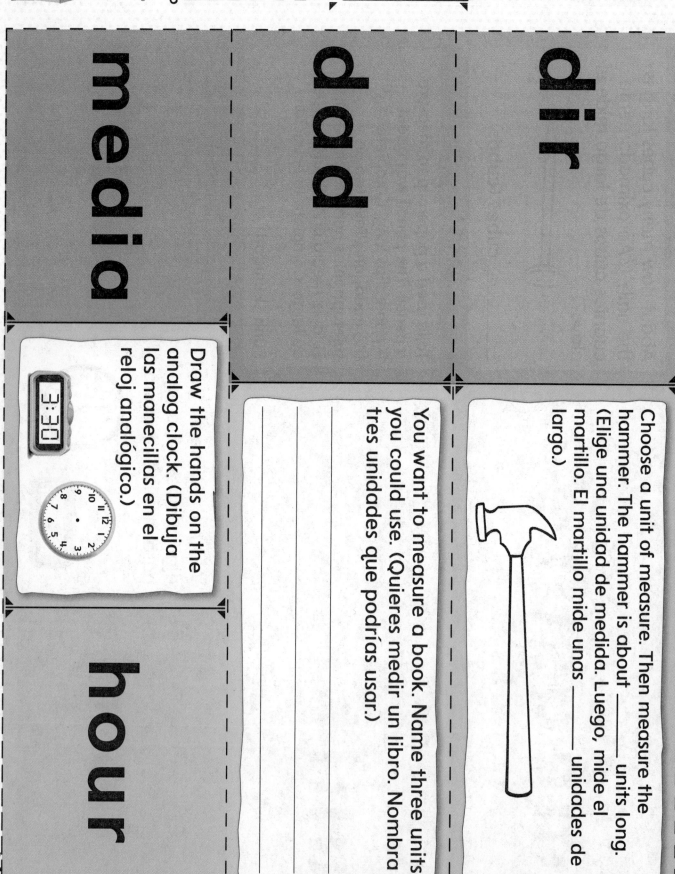

# Chapter 8

✂ cut on all dashed lines    fold on all solid lines

**reloj digital**

**analog**

What time is shown on the analog clock?
(¿Qué hora muestra el reloj analógico?)

\_\_\_\_\_ o'clock (en punto)

Chapter 8

cut on all dashed lines

fold on all solid lines

minute hand

horaria

Draw the minute hand to show 11:00. (Dibuja la manecilla de los minutos para mostrar las 11:00.)

Chapter 8

✂ cut on all dashed lines     ▱ fold on all solid lines

manecilla de los minutos

hour

Draw the hour hand to show 4:00. (Dibuja la manecilla horaria para mostrar las 4:00.)

VKV26  Chapter 8 Visual Kinesthetic Learning

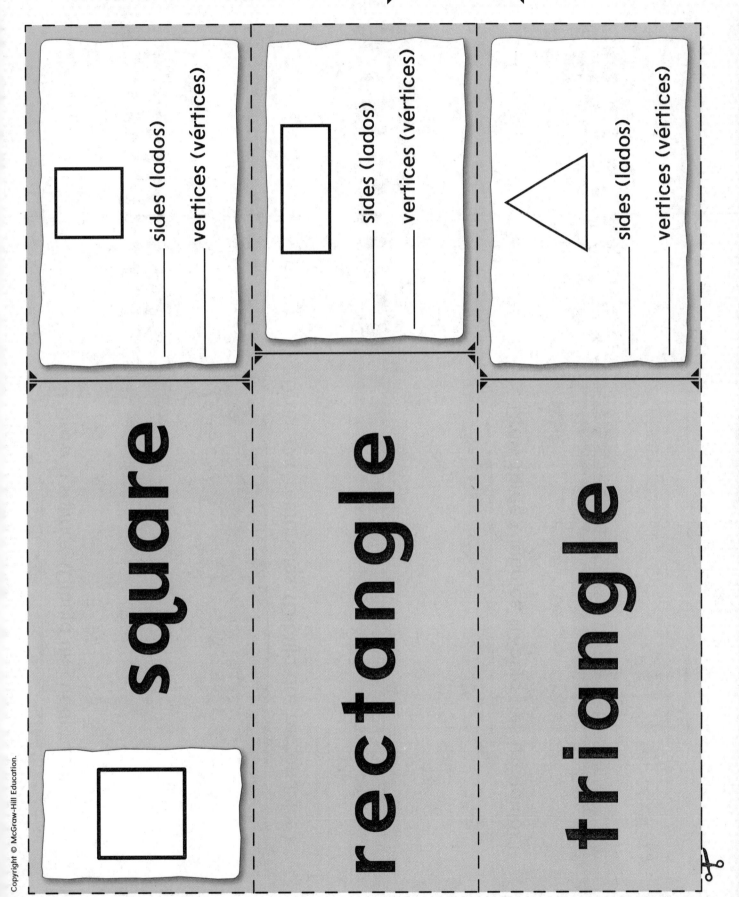

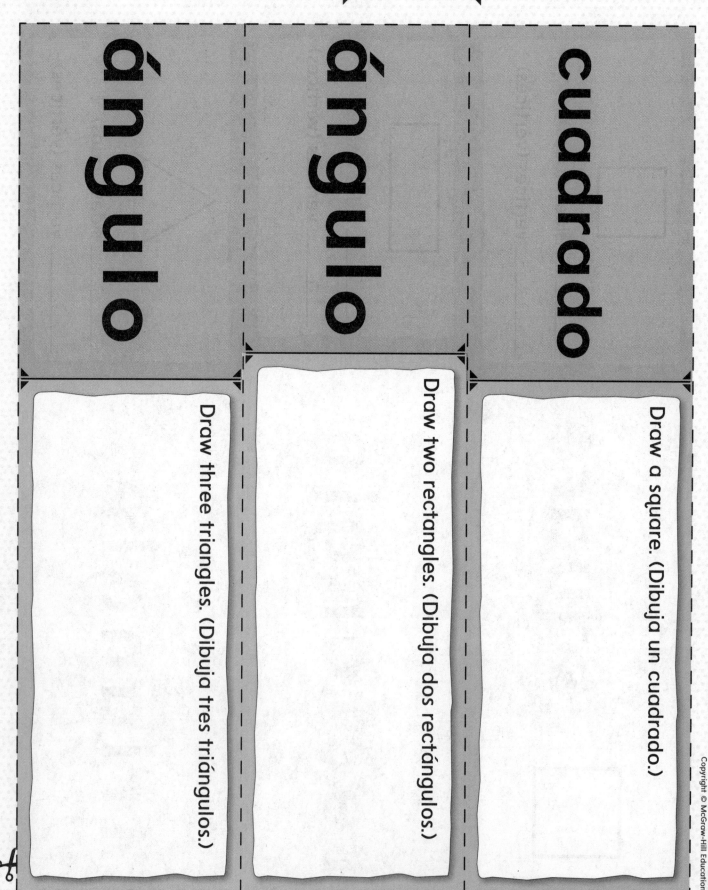

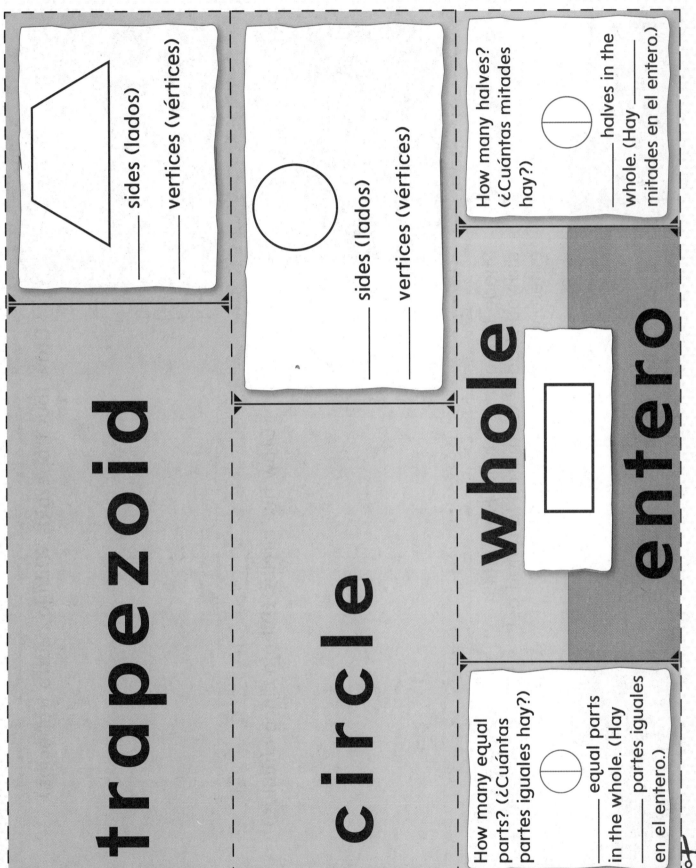

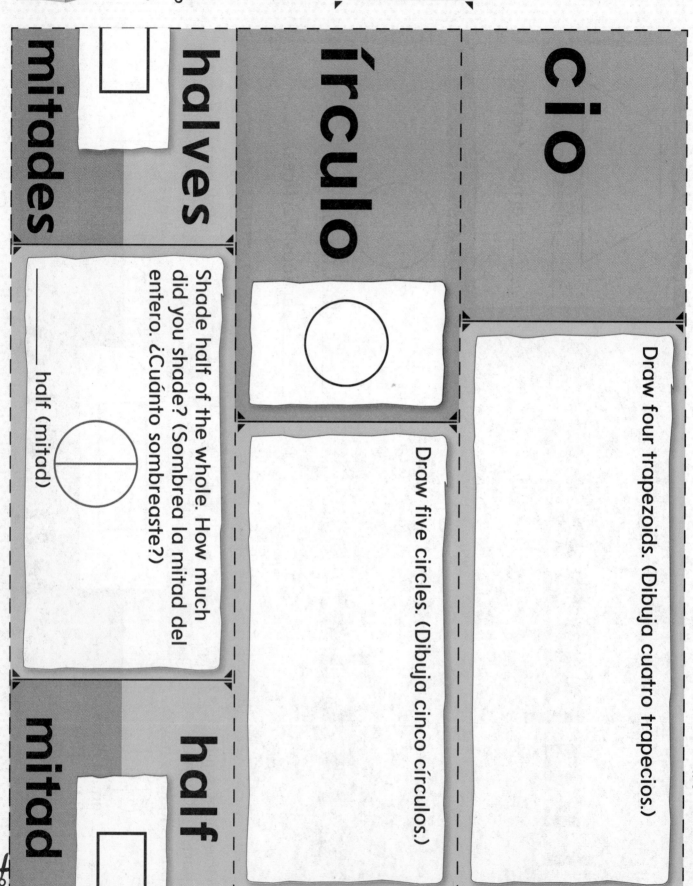

**Chapter 10**

cut on all dashed lines — fold on all solid lines

three-dimensional

rectangular

rectangular

___ faces (caras)
___ vertices (vértices)

bi

Which is two-dimensional?
(¿Cuál es bidimensional?)

Chapter 10 Visual Kinesthetic Learning  VKV31

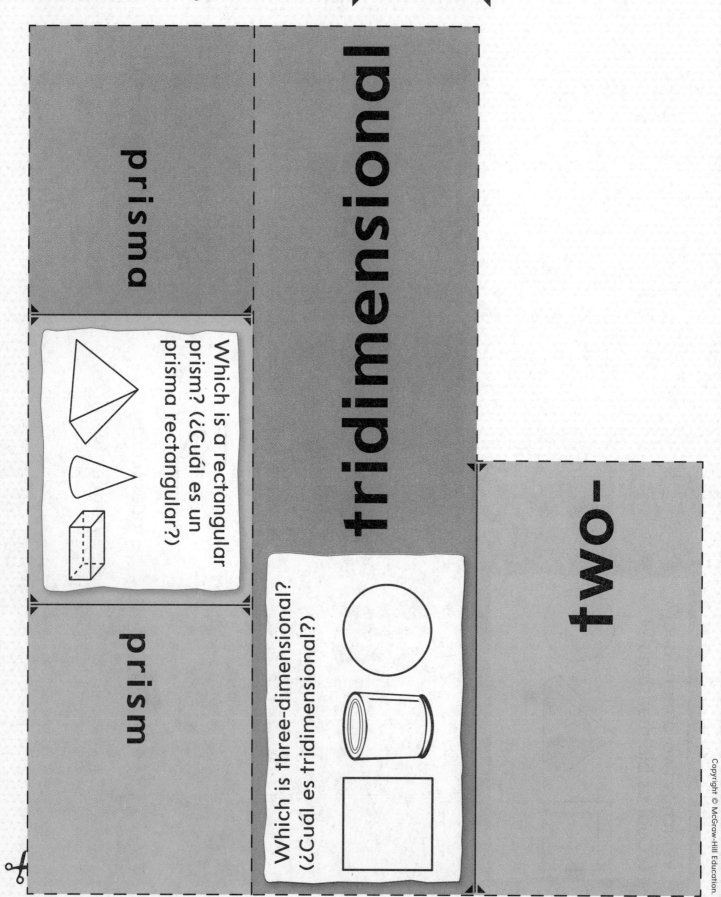

Chapter 10

cut on all dashed lines    fold on all solid lines

___ face (cara)
___ vertex (vértice)

___ faces (caras)
___ vertices (vértices)

cube

___ faces (caras)
___ vertices (vértices)

cylinder

Chapter 10 Visual Kinesthetic Learning   VKV33

Chapter 10

cut on all dashed lines  fold on all solid lines

ilindro

o

con

What shape are the faces on a cylinder? (¿Qué forma tienen las caras de un cilindro?)
square (cuadrado)
circle (círculo)
rectangle (rectángulo)

Circle the cubes. Underline the cones. (Encierra en un círculo los cubos. Subraya los conos.)

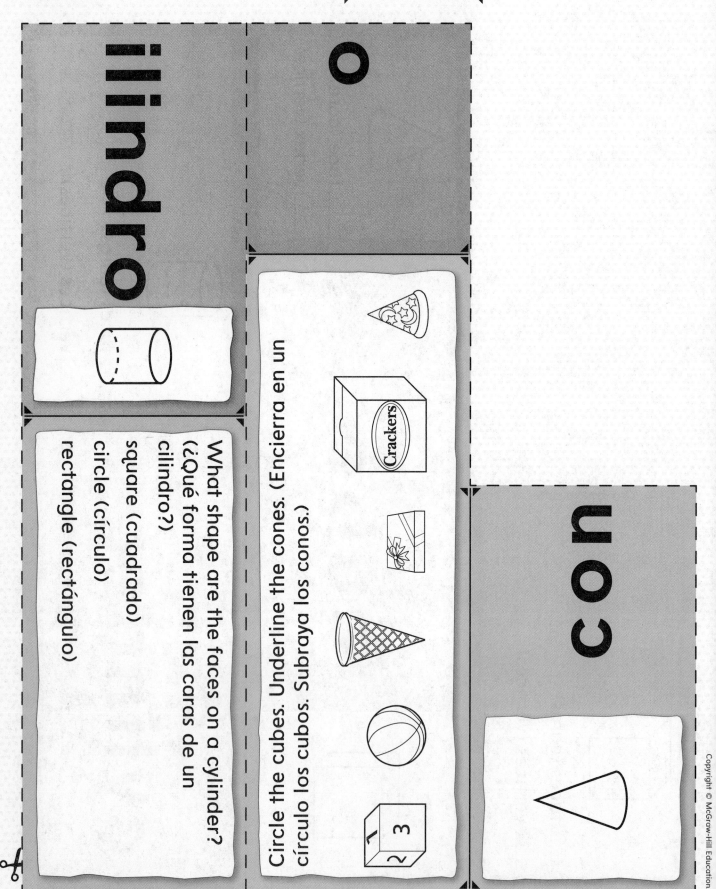